Global Warming

To my grandchildren, Daniel and Hannah
and their generation

Global Warming

THE COMPLETE BRIEFING

John Houghton

A LION BOOK

Text copyright © 1994 John Houghton

Published by
Lion Publishing plc
Sandy Lane West, Oxford, England
ISBN 0 7459 2458 1 (hardback)
ISBN 0 7459 3025 5 (paperback)
Lion Publishing
850 North Grove Avenue, Elgin, Illinois 60120, USA
ISBN 0 7459 2458 1 (hardback)
Albatross Books Pty Ltd
PO Box 320, Sutherland, NSW 2232, Australia
ISBN 0 7324 0843 1 (hardback)
ISBN 0 7324 0844 X (paperback)

First edition 1994

Acknowledgments
All figures and diagrams have been drawn
by Hardlines, Charlbury, except those on
the following pages: 12/13, 14, 22, 52, 54, 55,
60, 62, 63, 69, 74 (bottom), 83, 84, 85, 93, 158

A catalogue record for this book is available
from the British Library

Library of Congress CIP Data applied for

Printed and bound in Spain

CONTENTS

Introduction

Climate change and global warming are well up on the current political agenda. There are urgent questions everyone is asking: are human activities altering the climate? Is global warming a reality? How big are the changes likely to be? Will there be more serious disasters; will they be more frequent? Can we adapt to climate change or can we change the way we do things so that we can slow down the change or even prevent it occurring?

Because the Earth's climate system is highly complex, and because human behaviour and reaction to change is even more complex, providing answers to these questions is an enormous challenge to the world's scientists. As with many scientific problems only partial answers are available, but our knowledge is evolving rapidly, and the world's scientists have been addressing the problems with much energy and determination.

Three major pollution issues are often put together in people's minds: global warming, ozone depletion (the ozone hole) and acid rain. Although there are links between the science of these three issues (the chemicals which deplete ozone and the particles which are involved in the formation of acid rain also contribute to global warming), they are essentially three distinct problems. Their most important common feature is their large scale. In the case of acid rain the emissions of sulphur dioxide from one nation's territory can seriously affect the forests and the lakes of countries which may be downwind of the pollution. Global warming and ozone depletion are examples of global pollution—pollution in which the activities of one person or one nation can affect all people and all nations. It is only during the last thirty years or so that human activities have been of such a kind or on a sufficiently large scale that their effects can be significant globally. And because the problems are global, all nations have to be involved in their solution.

The key intergovernmental body which has been set up to assess the problem of global warming is the Intergovernmental Panel on Climate Change (IPCC), formed in 1988. At its first meeting in November of that year in Geneva, the Panel's first action was to ask for a scientific report so that, so far as they were known, the scientific facts about global warming could be established. It was imperative that politicians were given a solid scientific base from which to develop the requirements for action.

That first scientific report was published at the end of May 1990. On Monday 17 May I presented a preview of it to the then British Prime Minister, Mrs Margaret Thatcher, and members of her Cabinet at 10, Downing Street in London. I had been led to expect many interruptions and questions during my presentation. But the thirty or so Cabinet members and officials in the historic Cabinet room heard me in silence. They were clearly very interested in the report, and the questions and discussion afterwards demonstrated a large degree of concern for the world's environmental problems.

Since then the interest of many political leaders has been aroused—as has been shown by their attendance at two important world conferences concerned with global warming: the Second World Climate Conference in Geneva in 1990 and the United Nations Conference on Environment and Development (UNCED) in Rio de Janeiro in 1992. The Rio conference with over 25,000 people attending the main sessions and the many side meetings, was the largest conference ever held. Never before had a single conference seen so many of the world's leaders, and for that reason it is often referred to as the Earth Summit.

Much of the continuing assessment of climate change has been focused on the IPCC and its three working groups dealing respectively with science, impacts and response strategies. The IPCC's first report published in 1990 was a key input to the international negotiations which prepared the agenda for the UNCED Conference in Rio de Janeiro ; it was that IPCC assessment which provided much of the impetus for the Framework Convention on climate change signed at Rio by over 160 countries. As chairman or co-chairman of the Science Working Group I have been privileged to work closely with hundreds of scientific colleagues in many countries who readily gave of their time and expertise to contribute to the IPCC work.

For this book I have drawn heavily on the 1990 and 1992 reports of all three working groups of IPCC. Further, in putting forward options for action I have followed the logic of the Climate Convention. What I have said I believe to be consistent with the IPCC reports and with the implications of the Climate Convention. However, I must also emphasize that the choice of material and any particular views I put forward are entirely my own and should in no way be construed as the views of the IPCC.

During the preparation of both IPCC reports so far there has been considerable scientific debate about just how much can be said about likely climate change next century. Some researchers initially felt that the uncertainties were such that scientists should refrain from making any estimates or predictions for the future. However, it soon became clear that scientists have a responsibility to communicate the best possible information about the likely magnitude of climate change, along with clear statements of the assumptions made and the level of uncertainty in the estimates. Like weather forecasters, their results will not be entirely accurate, but can provide useful guidance.

Many books have been published on global warming. This book differs from the others because I have attempted to describe the science of global warming, its impacts and what action might be taken in a way which the intelligent non-scientist can understand. Although there are many numbers in the book—I believe the quantification of the problem to be very important—there are no mathematical equations. I have also used the minimum of jargon in the main text. Some technical explanations which would be of interest to the scientifically trained are included in some of the boxes. Others contain further material of specific interest.

I am grateful to many who have helped me with the provision and preparation of particular material for this book and to those who have read and helpfully commented on my drafts. There have been those who have been involved with the IPCC: Bert Bolin, the IPCC Chairman, Gylvan Meira Filho, my co-chairman on the IPCC Science Working Group, Robert Watson, co-chairman of the IPCC Working Group on Impacts and Response Strategies, Bruce Callander, Chris Folland, Niel Harris, Katherine Maskell, John Mitchell, Martin Parry, Peter Rowntree, Catherine Senior and Tom Wigley. Others I wish to thank are Myles Allen, David Carson, Jonathan Gregory, Donald Hay, David Fisk, Kathryn Francis, Michael Jefferson, Geoffrey Lean and John Twidell. The staff at Lion Publishing, Rebecca Winter, Nicholas Rous and Sarah Hall, have been most helpful in preparing the book for publication, especially in ensuring that it is as attractive and readable as possible. Finally, I owe an especial debt to my wife, Sheila, who gave me strong encouragement to write the book in the first place, and who has continued her encouragement and support through the long hours of its production.

1 Global Warming and Climate Change

*T*he phrase 'global warming' has become familiar recently as environmental issues have hit the headlines. Many opinions have been expressed concerning it, from the doom-laden to the dismissive. This book aims to state the current scientific position on global warming clearly, so that we can make informed decisions on the facts.

Is the climate changing?

In the year 2060 my grandchildren will be approaching seventy; what will their world be like? Indeed, what will it be like during the seventy years or so of their normal life span? Many new things have happened in the last seventy years which could not have been predicted in 1920. The pace of change is such that even more novelty can be expected in the next seventy. It is fairly certain that the world will be even more crowded and more connected. Will the increasing scale of human activities affect the environment? In particular, will the world be warmer? How is its climate likely to change?

Before studying future climate changes, what can be said about climate changes in the past? In the more distant past there have been very large changes. The last million years have seen a succession of major ice ages interspersed with warmer periods. The last of these ice ages began to come to an end about 20,000 years ago and we are now in what is called an interglacial period. Chapter 4 will focus on these times far back in the past. But have there been changes in the very much shorter period of living memory—over the past few decades?

Variations in day-to-day weather are occurring all the time; they are very much part of our lives. The climate of a region is its average weather over a period which may be a few months, a season or a few years. Variations in climate are also very familiar to us. We describe summers as wet or dry, winters as mild, cold or stormy. In the British Isles, as in many parts of the world, no season is the same as the last or indeed the same as any previous season, nor will it be repeated in detail next time round. Most of these variations we take for granted; they add a lot of interest to our lives. Those we particularly notice are the extreme situations and the climate disasters (for instance, Fig. 1.1 shows the significant climate events and disasters during the year 1991). Most of the worst disasters in the world are, in fact, weather- or climate-related. Table 1.1 lists them in order of severity although it does not include droughts, whose effects occur more slowly, but which are probably the most damaging disasters of all.

The 1980s: a remarkable decade

The 1980s have been unusually warm. Globally speaking, the decade has been the warmest since accurate records began somewhat over a hundred years ago and unusually warm years have continued into the 1990s. In terms of global average temperature, the year 1990 is the warmest on record and seven of the eight warmest years in the record have occurred in the 1980s and early 1990s.

The period has also been remarkable (just how remarkable will be considered later) for the frequency and intensity of extremes of weather and climate. For example, periods of unusually strong winds have been experienced in western Europe. During the early hours of the morning of 16 October 1987, over fifteen million trees were blown down in south-east England and the London area. The storm also hit Northern France, Belgium and Holland with ferocious intensity; it turned out to be the worst storm experienced in the area since 1703. Storm-force winds of similar intensity but covering a greater area of western Europe struck on several occasions in January and February 1990.

But those storms in Europe were mild by comparison with the much more intense and damaging storms other parts of the world have experienced during these years. About eighty hurricanes and typhoons—other names for tropical cyclones—occur around the tropical oceans each year, familiar enough to be given names. Hurricane Gilbert, which caused devastation on the island of Jamaica and the coast of Mexico in 1988, and Hurricane Andrew, which caused a great deal of damage in Florida and other regions of the southern United States in 1992, have been notable recent examples. Low-lying areas such as Bangladesh

TABLE 1.1 **Natural disasters 1947–1980 in order of severity**[1].

Natural disasters 1947–1980

	Type of Disaster	Deaths
1.	Tropical cyclones, hurricanes, typhoons	499,000
2.	Earthquakes	450,000
3.	Floods (other than associated with 1)	194,000
4.	Tornadoes and thunderstorms	29,000
5.	Snowstorms	10,000
6.	Volcanoes	9,000
7.	Heatwaves	7,000
8.	Avalanches	5,000
9.	Landslides	5,000
10.	Tidal waves (Tsunamis)	5,000

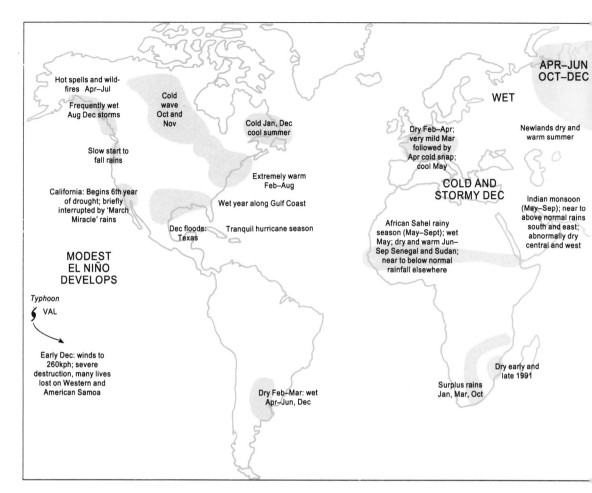

Hot spells and wild-fires Apr–Jul

Frequently wet Aug Dec storms

Slow start to fall rains

California: Begins 6th year of drought; briefly interrupted by 'March Miracle' rains

Cold wave Oct and Nov

Cold Jan, Dec cool summer

Extremely warm Feb–Aug

Wet year along Gulf Coast

Dec floods: Texas

Tranquil hurricane season

MODEST EL NIÑO DEVELOPS

Typhoon

⟨ VAL

Early Dec: winds to 260kph; severe destruction, many lives lost on Western and American Samoa

Dry Feb–Mar: wet Apr–Jun, Dec

Dry Feb–Apr; very mild Mar followed by Apr cold snap; cool May

COLD AND STORMY DEC

African Sahel rainy season (May–Sept); wet May; dry and warm Jun–Sep Senegal and Sudan; near to below normal rainfall elsewhere

APR–JUN OCT–DEC

WET

Newlands dry and warm summer

Indian monsoon (May–Sep); near to above normal rains south and east; abnormally dry central and west

Dry early and late 1991

Surplus rains Jan, Mar, Oct

FIG. 1.1 **Significant climate anomalies and events during 1991 as recorded by the Climate Analysis Centre of the United States[2].**

are particularly vulnerable to the storm surges associated with tropical cyclones; the combined effect of intensely low atmospheric pressure, extremely strong winds and high tides causes a surge of water which can reach far inland. In one of the worst such disasters this century over 250,000 people were drowned in Bangladesh in 1970. The people of that country experienced another storm of similar proportions in 1991 and smaller surges are a regular occurrence there.

The increase in storm intensity during recent years has been tracked

TABLE 1.2 **Losses (in thousand millions of US dollars adjusted to 1992 prices) in major windstorm catastrophes 1960–92 (mostly in North America and Europe), estimated by a research group advising the insurance industry[3].**

Losses in windstorm catastrophes

	Decade 1960–69	Decade 1970–79	Decade 1980–89	10 years 1983–92
Number of windstorm catastrophes	8	14	29	31
Economic losses	23	34	38	88
Insured losses	5	8	19	52

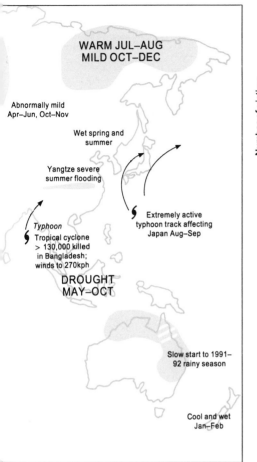

WARM JUL–AUG
MILD OCT–DEC

Abnormally mild
Apr–Jun, Oct–Nov

Wet spring and
summer

Yangtze severe
summer flooding

Typhoon
Tropical cyclone
> 130,000 killed
in Bangladesh;
winds to 270kph

Extremely active
typhoon track affecting
Japan Aug–Sep

DROUGHT
MAY–OCT

Slow start to 1991–
92 rainy season

Cool and wet
Jan–Feb

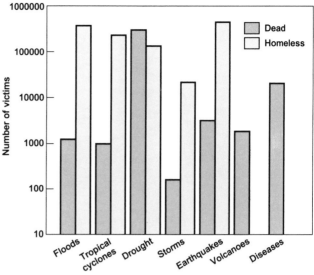

FIG. 1.2 **Recorded disasters in Africa, 1980–1989, estimated by the Organization for African Unity[4].**

by the insurance industry, which has been hit hard by recent disasters. Until the mid-1980s, it was widely thought that windstorms with insured losses exceeding one thousand million US dollars were only possible, if at all, in the United States. But the gales that hit western Europe in October 1987 heralded a series of windstorm disasters which make losses of 10 thousand million dollars seem commonplace. Hurricane Andrew, for instance, left in its wake insured losses estimated at 16 thousand million dollars. The estimates in Table 1.2 illustrate how the numbers and extent of such disasters have increased during the past three decades. The rate of economic loss has risen by a factor of four since the 1960s while the increase in insured losses is almost tenfold. Although some of this increase is due to the increased population over this period in particularly vulnerable areas, a large part of it seems to have arisen from the increased storminess in the late 1980s and early 1990s.

Windstorms are by no means the only weather and climate extremes that cause disasters. Floods due to unusually intense or prolonged rainfall or droughts because of long periods of reduced rainfall (or its complete absence) can be even more devastating to human life and property. These events occur frequently in many parts of the world especially in the tropics and sub-tropics. There have been notable examples during the last decade. In 1988, the highest flood levels ever recorded occurred in Bangladesh; 80 per cent of the entire country was affected. The Yangtze river region of China experienced a devastating flood in 1991. In 1993, flood waters rose to levels higher than ever recorded in the region of the

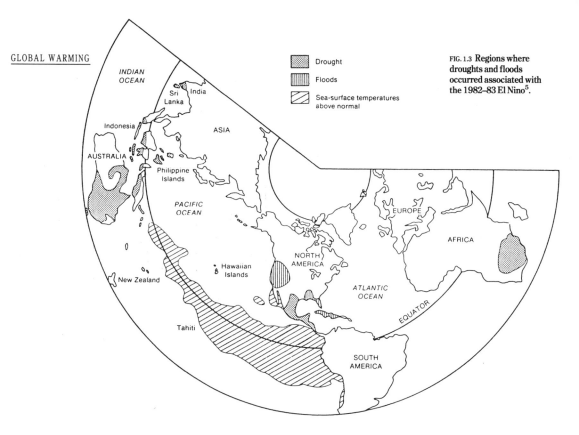

Drought

Floods

Sea-surface temperatures above normal

FIG. 1.3 **Regions where droughts and floods occurred associated with the 1982–83 El Nino[5].**

Mississippi and Missouri rivers in the United States, flooding an area equivalent in size to one of the Great Lakes. Large areas of Africa, both north and south, and of Australia have had droughts on a scale and for longer periods than any in living memory.

Because of the likely locations of floods and droughts, they often bear most heavily on the most vulnerable in the world, who can have little resilience to major disasters. Fig. 1.2 shows that climate-related disasters account for more than half of all disasters for the continent of Africa and illustrates the scale of the problem.

The El Nino

Rainfall patterns which lead to floods and droughts in tropical and semi-tropical areas are strongly influenced by the surface temperature of the

oceans around the world, particularly the pattern of ocean surface temperature in the Pacific off the coast of South America (see Chapter 5). About every three to five years a large area of warmer water appears and persists for a year or more. Because they usually occur around Christmas these are known as El Nino ('the boy child') events. They have been well known for centuries to the countries along the coast of South America because of their devastating effect on the fishing industry; the warm top waters of the ocean prevent the nutrients from lower, colder levels required by the fish from reaching the surface.

A particularly intense El Nino occurred in 1982–83; the anomalous highs in ocean surface temperature compared to the average reached 7°C. Droughts and floods somewhere in almost all the continents were

14

associated with that El Nino (Fig. 1.3). Like many events associated with weather and climate, El Ninos often differ very much in their detailed character. For instance, the El Nino event which began in 1990 and reached maturity early in 1992 weakened abruptly during mid-1992, but began to increase again early in 1993. The exceptional floods in the central United States and in the Andes, and the droughts in Australia and Africa, are probably linked with this unusually protracted El Nino. Studies with computer models of the kind described later in Chapter 5 provide a scientific basis for the links between the El Nino and these extreme weather events; they also give some confidence that useful forecasts of such disasters may one day be possible.

The effect of volcanic eruptions on temperature extremes

Volcanoes inject enormous quantities of dust and gases into the upper atmosphere. Large amounts of sulphur dioxide are included, which through photochemical reactions using the sun's energy are transformed to sulphuric acid and sulphate particles. Typically these particles remain in the stratosphere (the region of atmosphere above about 10km in altitude) for several years before they fall into the lower atmosphere and are quickly washed out by rainfall. During this period they disperse around the whole globe and cut out some of the radiation from the sun, thus tending to cool the lower atmosphere.

One of the largest volcanic eruptions this century was that from Mount Pinatubo in the Philippines on 12 June 1991 which injected about 20 million tons of sulphur dioxide into the stratosphere together with enormous amounts of dust. This stratospheric dust caused spectacular sunsets around the world for many months following the eruption. The amount of radiation from the sun reaching the lower atmosphere fell by about 2 per cent. Global average temperatures lower by about a quarter of a degree Celsius were experienced for the following two years. There is also evidence that some of the unusual weather patterns of 1991 and 1992, for instance unusually cold winters in the Middle East and mild winters in western Europe, were linked with effects of the volcanic dust.

Vulnerable to change

The occurrence of these extreme climate events and climate disasters has emphasized to all of us the importance of the climate to our lives and has caused countries around the world to realize their vulnerability to climate change—a vulnerability which is enhanced by rapidly increasing demands on resources.

But the question must be asked: how remarkable are these events? Do they point to a changing climate due to human activities? Do they provide evidence for global warming because of the increased carbon dioxide and other greenhouse gases being emitted into the atmosphere by burning fossil fuels?

Here a note of caution must be sounded. The range of normal natural climate variation is large. Climate extremes are nothing new. Climate records are continually being broken. In fact, a month without a broken record somewhere would itself be something of a record! Changes in climate which indicate a genuine long-term trend can only be

15

identified after many years.

Carbon dioxide in the atmosphere has been increasing over the past two hundred years and more substantially over the past fifty years. To identify climate change related to this carbon dioxide increase, we need to look for trends in global warming over similar lengths of time. They are long compared with both the memories of a generation and the period for which accurate and detailed records exist. Although, therefore, it can be ascertained that there has been more storminess, for instance, in the region of the north Atlantic during the late 1980s and early 1990s than there was in the previous two decades, it is not clear whether those years were that exceptional compared with other periods in the previous hundred years. There is even more difficulty in tracking detailed climate trends in many other parts of the world, owing to the lack of adequate records; further, trends in the frequency of rare events are very difficult to detect.

The generally cold period worldwide during the 1960s and early 1970s caused speculation that the world was heading for an ice age. A British television programme about climate change called *The ice age cometh* was prepared in the early 1970s and widely screened—but the cold trend soon came to an end. We must not be misled by our relatively short memories.

What is important is to continually make careful comparisons between practical observations of the climate and its changes and what scientific knowledge leads us to expect. During the last few years, as the occurrence of extreme events have made the public much more aware of environmental issues, scientists in their turn have become somewhat more sure about just what human activities are doing to the climate. Later chapters will look in detail at the science of global warming and at the climate changes that we can expect, as well as investigating how these changes fit in with the recent climate record. Here, however, is a brief outline of current scientific thinking on the problem.

The problem of global warming

Human industry and other activities such as deforestation are emitting increasing quantities of gases, in particular the gas carbon dioxide, into the atmosphere. Every year these emissions currently deliver over seven thousand million tons of carbon into the atmosphere, much of which is likely to remain there for a period of a hundred years or more. Because carbon dioxide is a good absorber of heat radiation coming from the Earth's surface, increased carbon dioxide acts like a blanket over the surface, keeping it warmer than it would otherwise be. With the increased temperature the amount of water vapour in the atmosphere also increases, providing more of a blanket effect and causing it to be even warmer.

Being kept warmer may sound appealing to those of us who live in cool climates. However, an increase in global temperature will lead to global climate change. If the change were small and occurred slowly enough we would almost certainly be able to adapt to it. However, with the rapid expansion taking place in the world's industry the change is unlikely to be either small or slow. The estimate I present in later chapters is that, in the absence of efforts to curb the rise in the emissions of carbon dioxide, the global average temperature will rise by about a quarter of a degree Celsius every ten years—or about two and a half degrees in a century.

This may not sound very much, especially when it is compared with normal temperature variations from day to night or between one day and the next. But it is not the temperature at one place but the temperature averaged over the whole globe. The predicted rate of change of two and a half degrees a century is probably faster than the global average temperature has changed at any time over the past ten thousand years. And as there is a difference in global average temperature of only about five or six degrees between the coldest part of an ice age and the warm period in between ice ages (see Fig. 4.4), we can see that a few degrees in this global average can represent a big change in climate.

Not all the climate changes will in the end be adverse. While some parts of the world experience more frequent or more severe droughts or floods, other parts perhaps in the sub-arctic may become more habitable. Even there, though, the likely rate of change will cause problems: large damage to buildings will occur in regions of melting permafrost, and trees in sub-arctic forests like trees elsewhere will need time to adapt to new climatic regimes.

Scientists are confident about the fact of global warming and climate change due to human activities. However, substantial uncertainty remains about just how large the warming will be and what will be the patterns of change in different parts of the world. Although some indications can be given, scientists cannot yet say with a lot of detail which regions will be most affected and in what way. Intensive research is needed to improve the confidence in scientific predictions.

Uncertainty and response

Until the predictions improve to the point where they can be used as a clear guide to action, politicians and others making decisions are faced with the need to weigh scientific uncertainty against the cost of the various actions which could be taken in response to the threat of climate change. Some action can be taken easily at relatively little cost (or even at a net saving of cost), for instance the development of programmes to conserve and save energy, and many schemes for reducing deforestation and encouraging the planting of trees. Other actions such as a large shift to renewable sources of energy (for example, biomass, hydro, wind or solar energy) in both the developed and the developing countries of the world will take some time. But here, too, a start can be made. What is important is that plans are made now in preparation for the major changes that will almost certainly be required.

In the following chapters I shall first explain the science of global warming, the evidence for it and the current state of the art regarding climate prediction. I shall then go on to say what is known about the likely impacts of climate change on human life—on water and food supplies for instance. The questions of why we should be concerned for the environment and what action should be taken in the face of scientific uncertainty is followed by consideration of the technical possibilities for large reductions in the emissions of carbon dioxide and how these might affect our energy sources and usage, including means of transport.

Finally I will address the issue of the 'global village'. So far as the environment is concerned, national boundaries are becoming less and less important; pollution in one country can now affect the whole world. And it is clear from our current scientific understanding that global warming poses a global challenge, which must be met by global solutions.

FOOTNOTES

1 After B.V. Shah, 1983, quoted in 'Natural Disaster Reduction: how meteorological services can help', *WMO* **No. 722**, 1989, World Meteorological Organization, Geneva.

2 From *World Climate News*, **No. 1**, June 1992: World Meteorological Organization, Geneva.

3 From G. Berz and K. Conrad, 'Winds of change', *The Review*, June 1993, pp 32–35.

4 From 'The role of the World Meteorological Organization in the International Decade for Natural Disaster Reduction' *WMO*, **No. 745**, 1990, World Meteorological Organization, Geneva.

5 Adapted from T.Y. Canby, 'El Nino's ill wind', *Natn. geogr. Mag.*, 1984, pp 144–83.

2 *The Greenhouse Effect*

The basic principle of global warming can be understood by considering the radiation energy from the sun which warms the Earth's surface and the thermal radiation from the Earth and the atmosphere which is radiated out to space. On average these two radiation streams must balance. If the balance is disturbed (for instance by an increase in atmospheric carbon dioxide) it can be restored by an increase in the Earth's surface temperature.

How the Earth keeps warm

To explain the processes which warm the Earth and its atmosphere, I will begin with a very simplified Earth. Suppose we could, all of a sudden, remove from the atmosphere all the clouds, the water vapour, the carbon dioxide and all the other minor gases and the dust leaving an atmosphere of nitrogen and oxygen only. Everything else remains the same. What, under these conditions, would happen to the atmospheric temperature?

The calculation is an easy one, involving a relatively simple radiation balance. Radiant energy from the sun falls on a surface of one square metre in area outside the atmosphere and directly facing the sun at a rate of about 1370 watts—about the power radiated by a reasonably sized domestic electric fire. However, few parts of the Earth's surface face the sun directly and in any case for half the time they are pointing away from the sun at night, so that the *average* energy falling on one square metre of a level surface outside the atmosphere is only one quarter of this[1] or about 343 watts. As this radiation passes through the atmosphere a small amount, about 6 per cent, is scattered back to space by atmospheric molecules. About 10 per cent on average is reflected back to space from the land and ocean surface. The remaining 84 per cent, or about 288 watts per square metre on average, remains actually to heat the surface—the power used by three good-sized electric light bulbs.

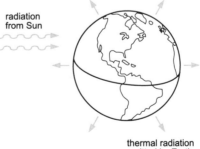

radiation from Sun

thermal radiation emitted by Earth

FIG. 2.1 **The radiation balance of planet Earth. The net incoming solar radiation is balanced by outgoing thermal radiation from the Earth.**

Atmospheric composition

Gas	Concentration: fraction* or parts per million by volume (ppmv)
Nitrogen (N_2)	0.78*
Oxygen (O_2)	0.21*
Water vapour (H_2O)	variable (0–0.02*)
Carbon dioxide (CO_2)	356
Methane (CH_4)	1.8
Nitrous Oxide (N_2O)	0.3
CFCs	0.001
Ozone (O_3)	variable (0-1,000)

TABLE 2.1 **The composition of the atmosphere, the main constituents (nitrogen and oxygen) and the greenhouse gases as in 1993.**

To balance this incoming energy, the Earth itself must radiate on average the same amount of energy back to space (Fig. 2.1) in the form of thermal radiation. All objects emit this kind of radiation; if they are hot enough we can see the radiation they emit. The sun at a temperature of about 6,000°C looks white; an electric fire at perhaps 800°C looks red. Cooler objects emit radiation which cannot be seen by our eyes and which lies at wavelengths beyond the red end of the spectrum—infrared radiation. On a clear, starry winter's night we are very aware of the cooling effect of this kind of radiation being emitted by the Earth's surface into space—it often leads to the formation of frost.

The amount of radiation emitted by the Earth's surface depends on its temperature—the warmer it is, the more radiation is emitted. The amount of radiation also depends on how absorbing the surface is; the greater the absorption, the more the radiation. Most of the surfaces on the Earth, including ice and snow, would appear 'black' if we could see them at infrared wavelengths; that means that they absorb nearly all the radiation which falls on them instead of reflecting it. It can be calculated[2] that, to balance the energy coming in, the average temperature of the Earth's surface must be −6°C to radiate the right amount[3]. This is much colder than is actually the case. In fact, an average of temperatures measured near the surface all over the Earth—over the oceans as well as over the land—averaging, too, over the whole year, comes to about 15°C. Some factor not yet taken into account is needed to explain this discrepancy.

The greenhouse effect

The gases nitrogen and oxygen which make up the bulk of the atmosphere (Table 2.1 gives details

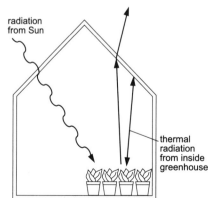

radiation from Sun

thermal radiation from inside greenhouse

FIG. 2.2 **A greenhouse has a similar effect to the atmosphere on the incoming solar radiation and the emitted thermal radiation.**

of the atmosphere's composition), neither absorb nor emit thermal radiation. It is the water vapour, carbon dioxide and some other minor gases present in the atmosphere in much smaller quantities (Table 2.1) which absorb some of the thermal radiation leaving the surface, acting as a partial blanket for this radiation and causing the difference of 21°C or so between the actual average surface temperature on the Earth of about 15°C and the figure of −6°C for an atmosphere containing nitrogen and oxygen only[4]. This blanketing is known as the *natural greenhouse effect* and the gases are known as greenhouse gases. It is called 'natural' because all the atmospheric gases (apart from the CFCs) were there long before human beings came on the scene. Later on I will mention the *enhanced greenhouse effect*: the added effect caused by the gases present in the atmosphere due to human activities such as the burning of fossil fuels and deforestation.

This is called the 'greenhouse effect' because the glass in a greenhouse possesses properties somewhat similar to our atmosphere (Fig. 2.2). Visible radiation from the sun passes almost unimpeded through the glass and is absorbed by the plants and the soil inside. The thermal radiation which is emitted by the plants and soil is, however, absorbed by the glass which re-emits some of it back into the greenhouse. The glass thus acts as a 'radiation blanket' helping to keep the greenhouse warm.

Before going further, it is interesting to mention some of those who have pioneered the science of the greenhouse effect[5]. The warming effect of the greenhouse gases in the atmosphere was first recognized in 1827 by the French scientist Jean-Baptiste Fourier, best known for his contributions to mathematics. He also pointed out the similarity between what happens in the atmosphere and in the glass of a greenhouse, which led to the name 'greenhouse effect'. The next step was taken by a British scientist, John Tyndall, who around 1860 measured the absorption of infrared radiation by carbon dioxide and water vapour; he also suggested that a cause of the ice ages might be a decrease in the greenhouse effect of carbon dioxide. It was a Swedish chemist, Svante Arrhenius, in 1896, who calculated the effect of an increasing concentration of greenhouse gases; he estimated that doubling the concentration of carbon dioxide would increase the global average temperature by 5 to 6°C, an estimate not too far from our present understanding. Nearly fifty years later, around 1940, G.S. Callendar, working in England, was the first to calculate the warming due to the increasing carbon dioxide from the burning of fossil fuels.

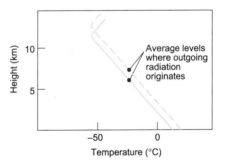

FIG. 2.3 **The distribution of temperature in a convective atmosphere (full line). The dotted line shows how the temperature increases when the amount of carbon dioxide present in the atmosphere is increased (in the diagram the difference between the lines is exaggerated—for instance, for doubled carbon dioxide in the absence of other effects the increase in temperature is about 1.2°C). Also shown for the two cases are the average levels from which thermal radiation leaving the atmosphere originates (about 6km for the unperturbed atmosphere).**

21

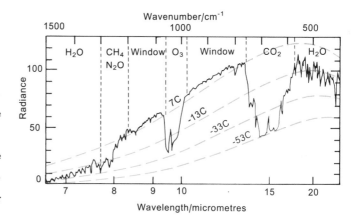

FIG. 2.4 Thermal radiation in the infrared (the visible part of the spectrum is between about 0.4 and 0.7 micrometres) emitted from the Earth's surface and atmosphere as observed over the Mediterranean Sea from a satellite instrument orbiting above the atmosphere, showing the parts of the spectrum where different gases contribute to the radiation[6]. Between wavelengths of about 8 and 14 micrometres, apart from the ozone band, the atmosphere, in the absence of clouds, is substantially transparent; this part of the spectrum is called a 'window' region. Superimposed on the spectrum are curves of radiation from a black-body at 7°C, −13°C, −33°C and −53°C. The units of radiance are watts per square metre per steradian per wavenumber.

The first expression of concern about the climate change which might be brought about by increasing greenhouse gases was in 1957, when Roger Revelle and Hans Suess of the Scripps Institute of Oceanography in California published a paper which pointed out that in the build-up of carbon dioxide in the atmosphere, human beings are carrying out a large-scale geophysical experiment. In the same year, routine measurements of carbon dioxide were started from the observatory on Mauna Kea in Hawaii. The rapidly increasing use of fossil fuels since then, together with growing interest in the environment, has led to the topic of global warming moving up the political agenda through the 1980s, and eventually to the Climate Convention signed in 1992—of which more in later chapters.

Back to the explanation: the transfer of radiation is only one of the ways heat is moved around in a greenhouse. The air inside also transfers heat by circulating. Some of the air movements are turbulent, mixing the heat up. Other heat transfer is due to less dense warm air moving upwards and more dense cold air moving downwards—a process called convection. Convective electric heaters used in the home heat

a room by stimulating convection in it. The situation in the greenhouse is therefore more complicated than would be the case if radiation was the only process of heat transfer.

Mixing and convection are also present in the atmosphere, although on a much larger scale, and in order to achieve a proper understanding of the greenhouse effect, convective heat transfer processes in the atmosphere must be taken into account as well as radiative ones.

Within the atmosphere itself (at least in the part of the atmosphere considered here) convection is, in fact, the dominant process for transferring heat. It acts as follows. The surface of the Earth is warmed by the sunlight it absorbs. Air close to the surface is heated and rises because of its lower density. As the air rises it expands and cools—just as the air cools as it comes out of the valve of a tyre. As some air masses rise, other air masses descend, so the air is continually turning over as different movements balance each other out—a situation of convective equilibrium. Temperature in the atmosphere falls with height at a rate determined by these convective processes; the drop turns out on average to be about 6°C per kilometre of height (Fig. 2.3).

A picture of the transfer of radiation in the atmosphere may be obtained by looking at the thermal radiation emitted by the Earth and its atmosphere as observed from instruments on satellites orbiting the Earth (Fig. 2.4). At some wavelengths in the infrared the atmosphere—in the absence of clouds—is largely transparent, just as it is in the visible part of the spectrum. If our eyes were sensitive at these wavelengths we would be able to see through the atmosphere to the sun, stars and moon above just as we can in the visible spectrum. At these wavelengths all the radiation originating from the Earth's surface leaves the atmosphere.

At other wavelengths radiation from the surface is strongly absorbed by some of the gases present in the atmosphere, in particular by water vapour and carbon dioxide.

Objects that are good absorbers of radiation are also good emitters of it. A black surface is both a good absorber and a good emitter, while a highly reflecting surface absorbs rather little and emits rather little too (which is why highly reflecting foil is used to cover the surface of a vacuum flask and why it is placed above the insulation in the lofts of houses).

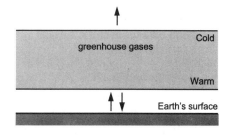

FIG. 2.5 **The blanketing effect of greenhouse gases.**

Absorbing gases in the atmosphere absorb some of the radiation emitted by the Earth's surface and in turn emit radiation out to space. The amount of thermal radiation they emit is dependent on their temperature.

Radiation is emitted out to space by these gases from levels somewhere near the top of the atmosphere—typically from between 5 and 10km high (see Fig. 2.3). Here, because of the convection processes mentioned earlier, the temperature is much colder—30 to 50°C or so colder—than at the surface. Because the gases are cold, they emit correspondingly less radiation. What these gases have done, therefore, is to absorb some of the radiation emitted by the surface but then to emit much less radiation out to space. The net loss of energy from the Earth's surface and the atmosphere is less than it would be if the absorbing gases were not present.

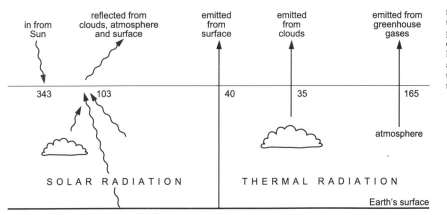

FIG. 2.6 **Components of the radiation (in watts per square metre) which on average enter and leave the Earth's atmosphere and make up the radiation budget for the atmosphere.**

They have, therefore, acted as a radiation blanket over the surface (note that the outer surface of a blanket is colder than inside the blanket) and helped to keep it warmer than it would otherwise be (Fig. 2.5).

There needs to be a balance between the radiation coming in and the radiation leaving the top of the atmosphere—as there was in the very simple model with which this chapter started. Fig. 2.6 shows the various components of the radiation entering and leaving the top of the atmosphere for the real atmosphere situation. On average, 240 watts per square metre of solar radiation is absorbed by the atmosphere and the surface; this is less than the 288 watts mentioned at the beginning of the chapter, because now the effect of clouds is being taken into account. Clouds reflect some of the incident radiation from the sun back out to space. However, they also absorb and emit thermal radiation and have a blanketing effect similar to that of the greenhouse gases. These two effects work in opposite senses: one (the reflection of solar radiation) tends to cool the Earth's surface and the other (the absorption of thermal radiation) to warm it. Careful consideration of these two effects shows that on average the net effect of clouds on the total budget of radiation results in a slight cooling of the Earth's surface[7].

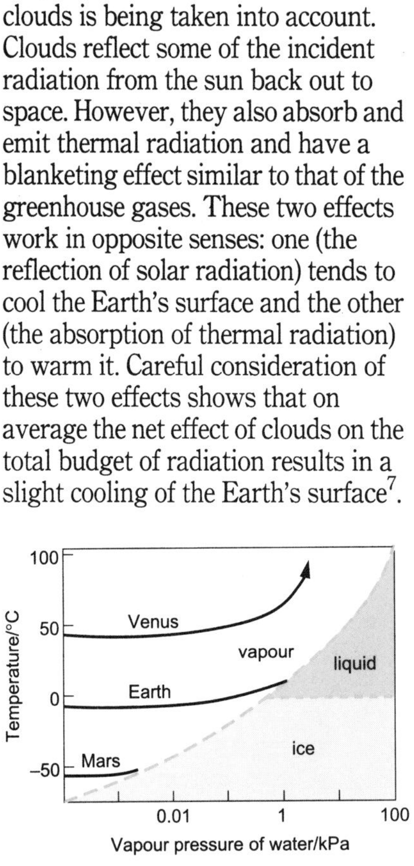

FIG. 2.7 Illustrating the evolution of the atmospheres of the Earth, Mars and Venus. In this diagram the surface temperatures of the three planets are plotted against the vapour pressure of water in their atmospheres as they evolved. Also on the diagram (dashed) are the phase lines for water, dividing the diagram into regions where vapour, liquid water or ice are in equilibrium. For Mars and the Earth the greenhouse effect is halted when water vapour is in equilibrium with ice or liquid water. For Venus no such halting occurs and the diagram illustrates the 'runaway' greenhouse effect[8].

The numbers in Fig. 2.6 demonstrate the required balance— 240 watts per square metre on average coming in and 240 watts per square metre on average going out. The temperature of the surface and hence of the atmosphere above adjusts itself to ensure that this balance is maintained. It is interesting to note that the greenhouse effect can only operate if there are colder temperatures in the higher atmosphere. Without the structure of decreasing temperature with height, therefore, there would be no greenhouse effect on the Earth.

Mars and Venus

Similar greenhouse effects also occur on our nearest planetary neighbours, Mars and Venus. Mars is smaller than the Earth and possesses, by Earth's standards, a very thin atmosphere. A barometer on the surface of Mars would record an atmospheric pressure less than 1 per cent of that on the Earth. Its atmosphere, however, consists almost entirely of carbon dioxide, which, as we have see, exerts a substantial greenhouse effect. Because Mars is 50 per cent further away from the sun than the Earth it receives less energy from the sun. If it had no atmosphere, to balance the incoming radiation from the sun its surface temperature would be about $-57°C$. In fact it is about $-47°C$; the greenhouse effect of the carbon dioxide atmosphere makes it some $10°C$ warmer.

The planet Venus, which can often be seen fairly close to the sun in the morning or evening sky, has a very different atmosphere to Mars. Venus is about the same size as the Earth. A barometer for use on Venus would need to survive very hostile

24

conditions. It would need to be able to measure a pressure about one hundred times as great as that on the Earth. Within the Venus atmosphere, which consists very largely of carbon dioxide, deep clouds consisting of droplets of almost pure sulphuric acid completely cover the planet and prevent most of the sunlight from reaching the surface. Some Russian space probes that have landed there have recorded what would be dusk-like conditions on the Earth—only 1 or 2 per cent of the sunlight present above the clouds penetrates that far. One might suppose, because of the small amount of solar energy available to keep the surface warm, that it would be rather cool. On the contrary; measurements from the same Russian space probes find a temperature there of about 525°C—a dull red heat, in fact.

The reason for this very high temperature is the greenhouse effect. Because of the very thick absorbing atmosphere of carbon dioxide, very little of the thermal radiation from the surface can get out. The atmosphere acts as such an effective radiation blanket that, although there is not much solar energy to warm the surface, the greenhouse effect amounts to nearly 500°C.

The 'runaway' greenhouse effect

What occurs on Venus is an example of what has been called the 'runaway' greenhouse effect. It can be explained by imagining the early history of the Venus atmosphere, which was formed by the release of gases from the interior of the planet. To start with it would contain a lot of water vapour, a powerful greenhouse gas (Fig. 2.7). The greenhouse effect of the water vapour would cause the temperature at the surface to rise.

The increased temperature would lead to more evaporation of water from the surface, giving more atmospheric water vapour, a larger greenhouse effect and therefore a further increased surface temperature. The process would continue until either the atmosphere became saturated with water vapour or all the available water had evaporated.

A runaway sequence something like this seems to have occurred on Venus. Why, we may ask, has it not happened on the Earth, a planet of about the same size as Venus and, so far as is known, of a similar initial chemical composition? The reason is that Venus is closer to the sun than the Earth; the amount of solar energy falling on Venus is about twice that falling on the Earth. The surface of Venus, when there was no atmosphere, would have started off at a temperature of just over 50°C (Fig. 2.7). Throughout the sequence described above for Venus, water on the surface would have been continuously boiling. Because of the high temperature, the atmosphere would never have become saturated with water vapour. The Earth, however, would have started at a colder temperature; at each stage of the sequence it would have arrived at an equilibrium between the surface and an atmosphere saturated with water vapour. There is no possibility of such runaway greenhouse conditions occurring on the Earth.

The enhanced greenhouse effect

After our excursion to Mars and Venus, let us return to Earth! The natural greenhouse effect is due to the gases water vapour and carbon dioxide present in the atmosphere in their natural abundances. The amount of water vapour in our atmosphere depends mostly on the temperature of the surface of the oceans; most of it originates through evaporation from the ocean surface and is not influenced directly by human activity. Carbon dioxide is different. Its amount has changed substantially—by about 25 per cent so far—since the Industrial Revolution, due to human industry and also because of the removal of forests (see Chapter 3). Future projections are that, in the absence of controlling factors, the rate of increase in atmospheric carbon dioxide will accelerate and that its atmospheric concentration will double from its pre-industrial value well within the next hundred years (Fig. 3.6).

This increased amount of carbon dioxide is leading to global warming of the Earth's surface because of its enhanced greenhouse effect. Let us imagine, for instance, that the amount of carbon dioxide in the atmosphere suddenly doubled, everything else remaining the same (Fig. 2.8). What would happen to the numbers in the radiation budget I presented earlier (Fig. 2.6)? The solar radiation budget would not be affected. The greater amount of carbon dioxide in the atmosphere means that the radiation emitted from it will originate on average from

a higher and colder level than before (Fig. 2.3). The thermal radiation budget will therefore be reduced, the amount of reduction being about 4 watts per square metre.

This causes a net imbalance in the overall budget of 4 watts per square metre. More energy is coming in than going out. To restore the balance the surface and atmosphere will warm up. If nothing changes apart from the temperature—in other words, the clouds, the water vapour, the ice and snow cover and so on, are all the same as before—the temperature change turns out to be about 1.2°C.

In reality, of course, many of these other factors will change, some of them in ways that add to the warming (these are called *positive feedbacks*), others in ways that might reduce the warming (*negative feedbacks*). The situation is therefore much more complicated than this simple calculation. These complications will be considered in more detail in Chapter 5. Suffice it to say here that the best estimate at the present time of the increased average temperature of the Earth's surface if carbon dioxide levels were to be doubled is about twice that of the simple calculation: 2.5°C. As the last chapter explained, for the global average temperature this is a *large* change. It is this global warming

FIG. 2.8 **Illustrating the enhanced greenhouse effect.** Under natural conditions (a) the net solar radiation coming in (S = 240 watts per square metre) is balanced by thermal radiation (L) leaving the top of the atmosphere; average surface temperature T_s is 15°C. If the carbon dioxide concentration is suddenly doubled (b), L is decreased by 4 watts per square metre. Balance is restored if nothing else changes (c) apart from the temperature of the surface and lower atmosphere which rises by 1.2°C. If feedbacks are also taken into account (d) the average temperature of the surface rises by about 2.5°C.

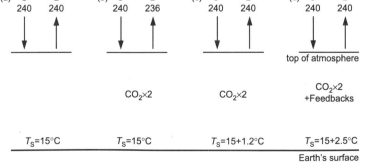

expected to result from the enhanced greenhouse effect which is the cause of current concern.

Having dealt with a doubling of the amount of carbon dioxide, it is interesting to ask what would happen if all the carbon dioxide were removed from the atmosphere. It is sometimes supposed that the outgoing radiation would be changed by 4 watts per square metre in the other direction and that the Earth would then cool by one or two degrees Celsius. In fact, that would happen if the carbon dioxide amount were to be halved. If it were to be removed altogether, the change in outgoing radiation would be around 25 watts per square metre—6 times as big—and the temperature change would be similarly increased. The reason for this is that with the amount of carbon dioxide currently present in the atmosphere there is maximum carbon dioxide absorption over much of the region of the spectrum where it absorbs (Fig. 2.4), so that a big change in gas concentration leads to a relatively small change in the amount of radiation it absorbs[9]. This is like the situation in a pool of water: when it is clear, a small amount of mud will make it appear muddy, but when it is muddy, adding more mud will make little difference.

An obvious question to ask is: has evidence of the enhanced greenhouse effect been seen in the recent climatic record? Chapter 4 will look at the record of temperature on the Earth during the last century or so, during which the Earth has warmed on average by about half a degree Celsius. However, because of natural climate variability this rise cannot yet be identified unequivocally as due to the enhanced greenhouse effect. If we cannot yet be positive that we have seen it, how can we be sure that the effect is real?

To summarize the argument so far:

■ No one doubts the reality of the natural greenhouse effect, which keeps us 21°C or so warmer than we would otherwise be. The science of it is well understood; it is similar science which applies to the enhanced greenhouse effect.

■ Substantial greenhouse effects occur on our nearest planetary neighbours, Mars and Venus. Given the conditions which exist on those planets, the sizes of their greenhouse effects can be calculated, and good agreement has been found with those measurements which are available.

■ The study of climates of the past gives some clues about the greenhouse effect, as Chapter 4 will show.

First, however, the greenhouse gases themselves must be considered. How does carbon dioxide get into the atmosphere, and what other gases affect global warming?

FOOTNOTES

1   It is one quarter because the area of the Earth's surface is four times the area of the disc which is the projection of the Earth facing the sun—see Fig. 2.1.

2   The radiation by a black body is the Stefan-Boltzmann constant (5.67×10^{-8} J m^{-2} K^{-4}s^{-1}) multiplied by the fourth power of the body's absolute temperature in degrees Kelvin. The absolute temperature is the temperature in degrees Celsius plus 273 (one degree K = one degree C).

3   These calculations using a simple model of an atmosphere containing nitrogen and oxygen only have been carried out to illustrate the effect of the other gases, especially water vapour and carbon dioxide. It is not, of course, a model that can exist in reality. All the water

vapour could not be removed from the atmosphere above a water or ice surface. Further, with an average surface temperature of – 6°C, in a real situation the surface would have much more ice cover. The additional ice would reflect more solar energy out to space leading to a further lowering of the surface temperature.

4   The above calculation is often carried out using a figure of 30 per cent for the average reflectivity of the Earth and atmosphere rather than the 16 per cent assumed here; the calculation of surface temperature then gives – 18°C for the average surface temperature rather than the – 6°C found here. The higher figure of 30 per cent for the Earth's average reflectivity is applicable when clouds are also included, in which case the average temperature of – 18°C is not applicable to the Earth's surface but to some appropriate level in the atmosphere. Further, clouds not only reflect solar radiation but also absorb thermal radiation, and so have a blanketing effect similar to greenhouse gases. For the purpose of illustrating the effect

of greenhouse gases, therefore, it is more correct to omit the effect of clouds from this initial calculation.

5   Further details can be found in J. Gribbin, *Hothouse Earth*, Bantam Press, 1990.

6   Spectrum taken with the infrared interferometer spectrometer flown on the satellite Nimbus 4 in 1971 and described by R.A. Hanel *et al.*, *Appl. Opt.* **10**, 1971, pp 1376–82.

7   More detail of the radiative effects of clouds is given in Chapter 5—see Figs 5.14 and 5.15.

8   From J.T. Houghton, *The Physics of Atmospheres*, second edition, CUP, 1986.

9   The dependence of the absorption on the concentration of the gas is approximately logarithmic.

3 *The Greenhouse Gases*

he greenhouse gases are those gases in the atmosphere which, by absorbing thermal radiation emitted by the Earth's surface, have a blanketing effect upon it. The most important of the greenhouse gases is water vapour, but its amount in the atmosphere is not changing directly because of human activities. The important greenhouse gases which are directly influenced by human activities are carbon dioxide, methane, nitrous oxide, the chlorofluorocarbons (CFCs) and ozone. This chapter will describe what is known about the origin of these gases, how their concentration in the atmosphere is changing and how it is controlled.

Which are the most important greenhouse gases?

Fig. 2.4 illustrated the regions of the infrared spectrum where the greenhouse gases absorb. Their importance as greenhouse gases depends both on their concentration in the atmosphere (Table 2.1) and on the strength of their absorption of infrared radiation. Both these quantities differ greatly for various gases.

Carbon dioxide is the most important of the greenhouse gases which are increasing in atmospheric concentration because of human activities. If we ignore the effects of the CFCs and of changes in ozone, which vary considerably over the globe and which are difficult to quantify, the increase in carbon dioxide (CO_2) has contributed about 70 per cent of the enhanced greenhouse effect to date, methane (CH_4) about 23 per cent, and nitrous

oxide (N_2O) about 7 per cent (Fig. 3.1).

Carbon dioxide and the carbon cycle.

Carbon dioxide provides the dominant means through which carbon is transferred in nature between a number of natural carbon reservoirs—a process known as the carbon cycle. We contribute to this

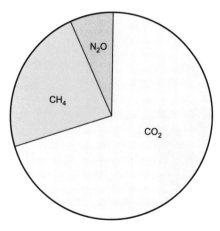

FIG. 3.1 **The contributions from the main greenhouse gases to the enhanced greenhouse effect from the beginning of the Industrial Revolution up to 1990.**

cycle every time we breathe. Using the oxygen we take in from the atmosphere, carbon from our food is burnt and turned into carbon dioxide which we then exhale; in this way we are provided with the energy we need to maintain our life. Animals contribute to atmospheric carbon dioxide in the same way; so do fires, rotting wood and decomposition of organic material in the soil and elsewhere. To offset these processes of respiration whereby carbon is turned into carbon dioxide, there are processes involving photosynthesis in plants and trees which work the opposite way; in the presence of light, they take in carbon dioxide, use the carbon for growth and let out the oxygen back into the atmosphere. Both respiration and photosynthesis also occur in the ocean.

Fig. 3.2 is a simple diagram of the way carbon cycles between the various reservoirs—the atmosphere, the oceans (including the ocean biota), the soil and the land biota

(biota is a word that covers all living things—plants, trees, animals and so on—on land and in the ocean, which make up a whole known as the biosphere). The diagram shows that the movements of carbon (in the form of carbon dioxide) into and out of the atmosphere are quite large; about one quarter of the total amount in the atmosphere is cycled in and out each year, half of this with the land biota and the other half through physical and chemical processes across the ocean surface. The land and ocean reservoirs are much larger than the amount in the atmosphere; small changes in these larger reservoirs could therefore have a large effect on the atmospheric concentration; the release of just 2 per cent of the carbon stored in the oceans would double the amount of atmospheric carbon dioxide.

Before human activities became a significant disturbance, and over periods short compared with geological timescales, the exchanges

FIG. 3.2 The reservoirs of carbon in the Earth, the biosphere, the ocean and the atmosphere and the annual exchanges of carbon dioxide (expressed in terms of mass of carbon it contains) between the reservoirs[1]. The units are thousand millions of tonnes or gigatonnes (Gt).

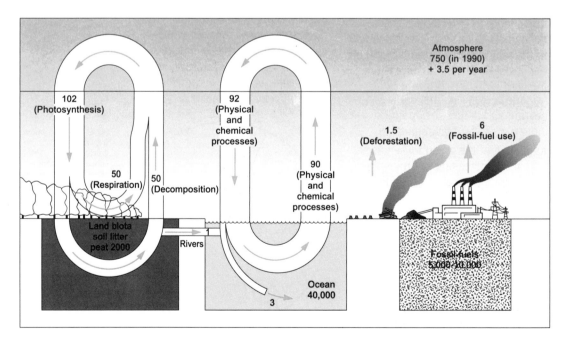

between the reservoirs were remarkably constant. For several thousand years before the beginning of industrialization around 1750, a steady balance was maintained, such that the concentration of carbon dioxide in the atmosphere as measured from ice cores (see Chapter 4) kept within about ten parts per million of a mean value of about 280 parts per million by volume (ppmv).

The Industrial Revolution disturbed this balance and has resulted in a concentration of carbon dioxide in the atmosphere which has increased by about 25 per cent, from 280 ppmv around the year 1700 to a value of over 350 ppmv at the present day (Fig. 3.3). Accurate measurements which have been made since 1959 from an observatory near the summit of Mauna Loa in Hawaii show that carbon dioxide is currently increasing each year by about 0.5 per cent or about 1.8 ppmv (Fig. 3.3b). This increase spread through the atmosphere adds about 3.8 thousand million tonnes (or gigatonnes, Gt) to the atmospheric carbon reservoir each year.

It is easy to establish how much coal, oil and gas are being burnt worldwide each year. Most of it is to provide energy for human needs: for heating and domestic appliances, for industry and for transport (considered in detail in Chapter 11). The amount of these fossil fuels burnt has increased rapidly since the Industrial Revolution (Fig. 3.4); currently the total contains about 6 Gt of carbon, nearly all of which enters the atmosphere as carbon dioxide. It is not so easy to estimate the amount of carbon dioxide released to the atmosphere because of changes of land use, in particular the burning and decay of forests.

Estimates vary widely; one is shown in Fig. 3.4a. Other estimates for current emissions from this source vary from about 0.5 to 2.5 Gt of carbon per year. Assuming a middle value of 1.5 Gt per year from changes in land use and deforestation and adding to it the 6 Gt per year from fossil fuel burning, the total entering the atmosphere from human activities sums to about 7.5 Gt per year. Since the annual net increase in the atmosphere is about 3.8 Gt, about half the 7.5 Gt of new carbon remains to increase the atmospheric

FIG. 3.3 (a) The increase of atmospheric carbon dioxide since 1700[2], showing measurements from ice cores in Antarctica (squares) and since 1957, direct measurements from the Mauna Loa observatory in Hawaii (triangles). (b) More detailed measurements of the increase of carbon dioxide since 1959 as observed at Mauna Loa (showing the annual cycle) and at the South Pole[3].

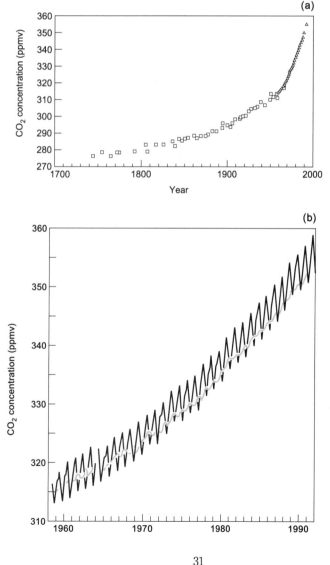

concentration. The other half is divided between the other two reservoirs: the oceans and the land biota.

About 95 per cent of fossil fuel burning occurs in the northern hemisphere, so there is more carbon dioxide there than in the southern hemisphere. The difference is currently about 2 parts per million (Fig. 3.3b) and, over the years, has grown in parallel with fossil fuel emissions, thus adding further compelling evidence that the atmospheric increase in carbon dioxide levels results from these emissions.

We turn now to what happens in the oceans. We know that carbon dioxide dissolves in water: carbonated drinks make use of that fact. Carbon dioxide is continually

The 'biological pump' in the oceans

In temperate and high latitudes there is a peak each spring in ocean biological activity. During the winter, water rich in nutrients is transferred from deep water to levels near the surface. As sunlight increases in the spring an explosive growth of the plankton population occurs, known as the 'spring bloom'. Pictures of the colour of the ocean taken from satellites orbiting the Earth can demonstrate dramatically where this is happening.

Plankton are small plants (phytoplankton) and animals (zooplankton) which live in the surface waters of the ocean; they range in size between about one thousandth of a millimetre across and the size of typical insects on land. Herbivorous zooplankton graze on phytoplankton; carnivorous zooplankton eat herbivorous zooplankton. Plant and animal debris from these living systems sinks in the ocean. While sinking some decomposes and returns to the water as nutrients, some (perhaps about 1 per cent) reaches the deep ocean or the ocean floor, where it is lost to the carbon cycle for hundreds, thousands or even millions of years. The net effect of the 'biological pump' is to reduce the amount of carbon in the surface waters, enabling the drawdown of more carbon dioxide from the atmosphere in order to restore the surface equilibrium.

Evidence of the importance of the 'biological pump' comes from the paleoclimate record from ice cores (see Chapter 4). One of the constituents from the atmosphere trapped in bubbles in the ice is the gas methyl sulphonic acid which originates from decaying ocean plankton; its concentration is therefore an indicator of plankton activity. As the global temperature began to increase when the last ice age receded nearly 20,000 years ago and as the carbon dioxide in the atmosphere began to increase (Fig. 3.3), the methyl sulphonic acid concentration decreased. An interesting link is thereby provided between the carbon dioxide in the atmosphere and marine biological activity. During the cold periods of the ice ages, enhanced biological activity in the ocean could have been responsible for maintaining the atmospheric carbon dioxide at a lower level of concentration— the 'biological pump' was having an effect.

The question then remains as to why the ice ages should be periods of greater marine biological activity than the warm periods in between. A British oceanographer, Professor John Woods, has suggested that the key may lie in what happens in the winter as nutrients are fed into the upper ocean ready for the spring bloom. When there is less atmospheric carbon dioxide, the cooling by radiation from the surface of the ocean increases. Since convection in the upper layers of the ocean is driven by cooling at the surface, the increased cooling results in a greater depth of the mixed layer near the top of the ocean where all the biological activity occurs. This is an example of a positive biological feedback; a greater depth of layer means more plankton growth. Woods calls it the 'plankton multiplier'[4].

being exchanged with the air above the ocean across the whole ocean surface (about 90 Gt per year is so exchanged), particularly as waves break. An equilibrium is set up between the concentration of carbon dioxide dissolved in the surface waters and the concentration in the air above the surface. The chemical laws governing this equilibrium are such that if the atmospheric concentration changes by 10 per cent the concentration in solution in the water changes by only one tenth of this: 1 per cent.

This change will occur quite rapidly in the upper waters of the ocean, the top hundred metres or so, but exchange with the lower levels in the ocean takes longer. Access from the surface to most parts of the very deep ocean takes several hundred to over a thousand years. So the oceans do not provide as immediate a sink for increased atmospheric carbon dioxide as might be suggested by the size of the exchanges with the large ocean reservoir. For short-term changes only the surface layers of water play a large part in the carbon cycle.

Biological activity in the oceans also plays an important role. It may not be immediately apparent, but the oceans are literally teeming with life. Although the total mass of living matter within the oceans is not large, it has a high rate of turnover. Living material in the oceans is produced at some 30–40 per cent of the rate of production on land. Most of this production is of plant and animal plankton which go through a rapid series of life cycles. As they die and decay some of the carbon they contain is carried downwards into deep water or to the ocean bottom where, so far as the carbon cycle is

concerned, it is out of circulation for hundreds or thousands of years. This process, whose contribution to the carbon cycle is known as the 'biological pump' (see box), has been important in determining the changes of carbon dioxide concentration in both the atmosphere and the ocean during the ice ages described in the last chapter. However, its influence on time-scales

FIG. 3.4 **(a) The annual growth of the amount of carbon dioxide in the atmosphere (Atm) together with estimates of the emissions from the burning of fossil fuels (FF), and from changes in land use and deforestation (Def) which sum to give the total emissions (Tot Em)[5].**

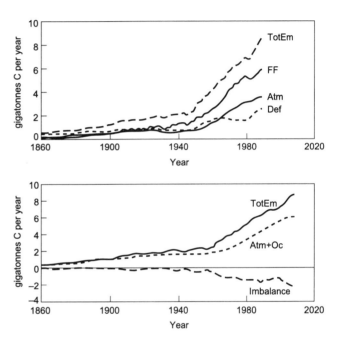

less than a few centuries is small, so its contribution so far to the sequestration (removal) of the extra carbon in the atmosphere due to human activities this century can be neglected.

Computer models—which calculate solutions for the mathematical equations describing a given physical situation, in order to predict its behaviour—have been set up to describe in detail the exchanges of carbon between the atmosphere and different parts of the ocean. To test the

(b) The total emissions of carbon dioxide generated by human activity as in (a) (Tot Em), the estimated amount entering the atmosphere and the ocean (Atm + Oc) and the imbalance[5].

validity of these models, they have also been applied to the dispersal in the ocean of the carbon isotope ^{14}C which entered the ocean after the nuclear tests of the 1950s, which they describe quite well. From the model results, it is estimated that between 1.5 and 2.5 Gt of the carbon dioxide added to the atmosphere each year ends up in the oceans. Observations of the relative distribution of the other isotopes of carbon in the atmosphere and in the oceans also confirm this estimate (see box).

As was seen earlier, about half the 7.5 Gt of carbon which enters the atmosphere each year as carbon dioxide from human-generated sources remains in the atmosphere. If about 2 Gt of the rest ends up in the oceans, there remains an imbalance of about 1.7 Gt or so to be accounted

for. An estimate of the size of this imbalance over the last 50 years or so is shown in Fig. 3.4b.

This imbalance may partially arise because of the uncertainty in the estimates that make up the budget. For instance, it may be that the carbon dioxide released to the atmosphere from land use changes has been poorly estimated; a value closer to the lower estimate of 0.5 Gt per year rather than 1.5 Gt may be nearer to the truth. But all of the imbalance cannot be accounted for in this way. Some must result from other changes in the land biosphere which have not yet been well quantified.

A clue to the origin of the imbalance is provided from observations of the atmospheric concentration of carbon dioxide

What we can learn from carbon isotopes

Isotopes are forms of one element with different atomic weights. Three isotopes of carbon are important in studies of the carbon cycle: the most abundant isotope ^{12}C which makes up 98.9 per cent of ordinary carbon, ^{13}C present at about 1.1 per cent and the radioactive isotope ^{14}C which is present only in very small quantities. About ten kilograms of ^{14}C are produced in the atmosphere each year by the action of particle radiation from the sun; half of this will decay into nitrogen over a period of 5,730 years (the 'half-life' of ^{14}C).

When carbon in carbon dioxide is taken up by plants and other living things, less ^{13}C is taken up in proportion than ^{12}C. Fossil fuel such as coal and oil was originally living matter so also contains less ^{13}C (by about 18 parts per thousand) than the carbon dioxide in ordinary air in the atmosphere today. Adding carbon to the atmosphere from burning forests, decaying vegetation or fossil fuel will therefore tend to reduce the proportion of ^{13}C.

Because fossil fuel has been stored in the Earth for much longer than 5,730 years (the half-life of ^{14}C) it contains no ^{14}C at all. As carbon from fossil fuel is added to the atmosphere, therefore, the proportion of ^{14}C in the atmosphere is also reduced.

By studying the ratio of the different isotopes of carbon in the atmosphere, in the oceans, in gas trapped in ice cores and in tree rings, it is possible to find out where the additional carbon dioxide in the atmosphere has come from and also what amount has been transferred to the ocean. For instance, it has been possible to estimate for different times how much carbon dioxide has entered the atmosphere from the burning or decay of forests and other vegetation and how much from fossil fuels.

Similar isotopic measurements on the carbon in atmospheric methane provide information about how much methane from fossil fuel sources has entered the atmosphere at different times.

which, each year, show a regular cycle (Fig. 3.3b). Carbon dioxide is removed from the atmosphere during the growing season and returned as the vegetation dies away in the

winter. Since there is more vegetation growth in the northern hemisphere than the southern, a minimum in the annual cycle of carbon dioxide in the atmosphere occurs in the northern summer. Detailed studies of this annual cycle together with the variations of carbon dioxide concentration between the hemispheres indicate that much of the imbalance must lie in the northern hemisphere.

What are the possible processes in the land biosphere which might account for this imbalance? One is the carbon dioxide fertilization effect; it is well established that increased carbon dioxide, under appropriate conditions, leads to increased growth (see box in Chapter 7). But it is not known to what extent the effect leads to a net increase in the total amount of carbon in the plants, the trees and the soils. Another possible process is the increasing use of fertilizers, which may be leading to more overall fixing of carbon in the world's agriculture. Yet another contribution to the imbalance may come from the regrowth of forests, especially at mid-latitudes. What is clear is that the processes involved in the carbon cycle are complex and vary considerably from place to place over the globe. Better understanding and quantification of all the processes involved is urgently required[7].

The carbon dioxide fertilization effect is an example of a biological feedback process. It is a negative feedback because, as carbon dioxide increases, it tends to increase the take-up of carbon dioxide by plants and therefore to reduce the amount in the atmosphere, decreasing the rate of global warming. Positive feedback processes, which would tend to accelerate the rate of global warming,

Feedbacks in the biosphere

As the greenhouse gases carbon dioxide and methane increase in concentration in the atmosphere, biological or other feedback processes are those processes occurring in the biosphere which, as a result of the increase, would tend to increase further the gas concentration (a positive feedback) or to decrease it (a negative feedback).

Two feedbacks, one positive (the plankton multiplier in the ocean) and one negative (carbon dioxide fertilization) have already been mentioned in the text. Three other positive feedbacks are potentially important, although our knowledge is currently insufficient to quantify them at all precisely.

One is the effect of higher temperatures on respiration, especially of microbes in soils, leading to increased carbon dioxide emissions. A second is the reduction of growth or the die-back especially in forests because of the stress caused by climate change. The third positive feedback is the release of methane, as temperatures increase— from wetlands and from very large reservoirs of methane trapped in sediments in a hydrate form (tied to water molecules when under pressure)—mostly at high latitudes. Methane has been generated from the decomposition of organic matter present in these sediments over many millions of years. Because of the depth of the sediments this latter feedback is unlikely to become operative to a significant extent during the next century. However, were global warming to continue to increase substantially for more than a hundred years, it is estimated that releases from hydrates could become the largest single contributor of methane emissions into the atmosphere.[8]

TABLE 3.1 **Assumptions regarding emissions from different sources for IPCC scenario IS 92a.** The emissions labelled 'energy' include those from cement production (about 0.2 Gt in 1990).

also exist; in fact there are more potentially positive processes than negative ones (see box). Although scientific knowledge cannot yet put precise figures on them, some of the positive feedbacks could be large, especially if carbon dioxide were to continue to increase, with its associated global warming, through the twenty-first century into the twenty-second.

The uptake by the oceans and by the land biota of about half the carbon dioxide added to the atmosphere both occur quite rapidly. What happens to the other half? It will eventually be taken up in the other reservoirs, particularly in the ocean. It cannot, however, be absorbed to any significant extent in the top part of the ocean. Getting it down to the deep ocean takes time of the order of several hundred years— the time for the top water to mix with water at lower levels.

This long time for the concentration of carbon dioxide in the atmosphere to respond to change has important implications. Suppose, for instance, that all emissions into the atmosphere from human activities were suddenly halted. No immediate change would occur in the atmospheric concentration, which would decline only slowly. We could not expect it to approach its pre-industrial value for several hundred years.

But emissions of carbon dioxide are not halting, nor are they slowing; their increase is, in fact, becoming larger each year. Later chapters will present estimates of climate change next century due to the increase in greenhouse gases. A prerequisite for such estimates is the knowledge of what changes in carbon dioxide emissions are likely to be. Guessing what will happen in the future is, of course, not easy. Because nearly everything we do has an influence on the emissions of carbon dioxide, it means estimating how human beings will behave and what their activities are likely to be. For instance, assumptions need to be made about population growth, economic growth,

Projected emissions of carbon dioxide from different sources (in Gt of carbon)

	1990	2025	2050	2100
Energy	6.2	11.1	13.7	20.4
Deforestation	1.3	1.1	0.8	-0.1
Total	7.5	12.2	14.5	20.3

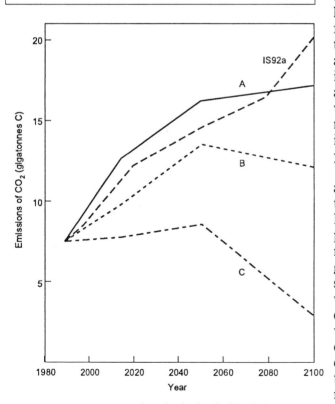

FIG. 3.5 **Scenarios of carbon dioxide emissions next century from energy generation for IPCC scenario IS92a and WEC scenarios A, B and C (details in Table 11.2). The emissions from WEC scenario B1 are not plotted separately; up to the year 2020 they are almost identical with those from IS 92a.**

energy use, the development of energy sources and the likely influence of pressures to preserve the environment. These assumptions are required for all countries of the world, both developing as well as developed ones. Further, since any assumptions made are unlikely to be fulfilled accurately in practice, it is necessary to make a variety of different assumptions, so that we can get some idea of the range of possibilities. Such possible futures are called *scenarios*.

The carbon dioxide emissions which would result from different scenarios prepared by the Intergovernmental Panel on Climate Change (IPCC)[9] and the World Energy Council (WEC)[10] are shown in Figure 3.5. The IPCC scenario IS 92a assumes a world population a little over double its present level by 2100, moderate economic growth and no strong pressure to reduce carbon dioxide emissions for environmental reasons. In the absence of strong environmental action it can perhaps be considered to be the most likely scenario—the 'business-as-usual' scenario. It shows close to a threefold rise in total emissions during next century (Table 3.1), made up of a large increase from the energy sector and a decrease of emissions from deforestation as the area of forest left in the world diminishes. Details of the assumptions regarding the emissions from energy production for the WEC scenarios are given in Chapter 11 (see Table 11.2).

To turn the emission scenarios into future projections of atmospheric carbon dioxide concentrations, they need to be incorporated into a computer model (see also Chapter 5) of the carbon cycle which includes descriptions of

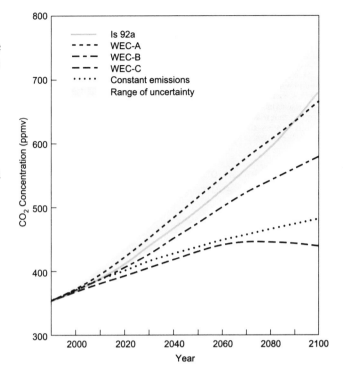

all the exchanges already mentioned. A number of scientists in different countries have developed carbon cycle models. Fig. 3.6 shows the results from the application of one such model developed by Professor Tom Wigley at the Climatic Research unit of the University of East Anglia to a number of different emission scenarios[11]. Other models give similar results. Fig. 3.6 also provides an indication of the uncertainty in the model arising from inadequate understanding of the carbon cycle, and in particular of the reason for the imbalance which currently exists in balancing the carbon cycle budget.

The information in Fig. 3.5 and 3.6 provides essential input for our discussion in later chapters of the consequences of increasing greenhouse gas emissions. Chapter 6 looks at the likely climate change for the IS 92a 'business-as-usual' scenario; later on Chapter 10

FIG. 3.6 Concentrations of carbon dioxide in the atmosphere resulting from the emission scenarios of Fig. 3.5[12] together with emissions from deforestation (Table 3.1). Also shown is the concentration curve for a scenario in which the emissions of carbon dioxide from fossil fuel emissions are kept constant beyond the year 2000 at 7.1 Gt of carbon. The coloured area around the curve for the IS 92a scenario provides an estimate of the uncertainty in the model computations. Similar uncertainty applies to the other scenarios.

considers the changes which may occur under some of the other scenarios. The WEC scenario C is particularly interesting, as it is only under that scenario that the increase in carbon dioxide concentration in the atmosphere is halted before the end of the next century and a stable level is reached.

Other greenhouse gases

METHANE

Methane is the main component of natural gas. Its common name used to be marsh gas because it can be seen bubbling up from marshy areas where organic material is decomposing. Data from ice cores show that for at least two thousand years before 1800 its concentration in the atmosphere was about 0.8 ppmv. Since then its concentration has more than doubled and is increasing on

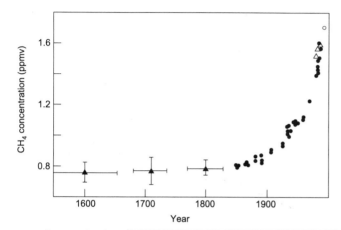

FIG. 3.7 **Concentration of methane in the atmosphere over the past few centuries as measured from air trapped in ice cores[13]. The value (full circle) for 1992 is also plotted.**

Sources and sinks of methane

Sources

NATURAL:		
Wetlands	115	(100–200)
Termites	20	(10–50)
Ocean	10	(5–20)
Freshwater	5	(1–25)
Methane hydrate	5	(0–5)
HUMAN-GENERATED:		
Coal mining, natural gas, petroleum industry	100	(70–120)
Rice paddies	60	(20–150)
Enteric fermentation	80	(65–100)
Animal wastes	25	(20–30)
Domestic sewage treatment	25	?
Landfills	30	(20–70)
Biomass burning	40	(20–80)

Sinks

Atmospheric removal	470	(420–520)
Removal by soils	30	(15–45)
Atmospheric increase	32	(28–37)

TABLE 3.2 **Estimated sources and sinks of methane in millions of tonnes per year[14]. The first column shows the best estimate from each source; the second column illustrates the uncertainty in the estimates by giving a range of values.**

average at about 1 per cent per year (Fig. 3.7)[15]. Although the concentration of methane in the atmosphere is much less than that of carbon dioxide (less than 2ppmv compared with about 350ppmv for carbon dioxide), its greenhouse effect is far from negligible. That is because the enhanced greenhouse effect caused by a molecule of methane is about 7.5 times that of a molecule of carbon dioxide[16].

The main natural source of methane is from wetlands. A variety of other sources result directly or indirectly from human activities, for instance from leakage from natural gas pipelines and from oil wells, from generation in rice paddy fields, from enteric fermentation (belching) from cattle and other livestock, from the decay of rubbish in landfill sites and from wood and peat burning. Details of the best estimates of the sizes of these sources are shown in Table 3.2. Attached to many of the numbers is a wide range of uncertainty. It is, for instance, difficult to estimate the amount produced in paddy fields averaged on a worldwide basis. The amount varies enormously during the rice growing season and also very widely from region to region. It can easily be imagined that similar problems arise when trying to estimate the amount produced by animals. Measurements of the proportions of the different isotopes of carbon (see box on p 34) in atmospheric methane assist considerably in helping to tie down the proportion which comes from fossil fuel sources, such as leakage from mines and from natural gas pipelines.

The main loss process for methane from the atmosphere is through chemical destruction. It reacts with hydroxyl (OH) radicals, which are present in the atmosphere because of processes involving sunlight, oxygen, ozone and water vapour. The average lifetime of methane in the atmosphere is determined by the rate of this loss process. At about 11 years, it is much shorter than the lifetime of carbon dioxide.

Although methane sources cannot be pinned down very precisely, most of the sources apart from natural wetlands are closely associated with human activities. It is interesting to note that the increase of atmospheric methane (Fig. 3.7) follows very closely the growth of human population since the Industrial Revolution. If no attempt is made to control human-related sources of methane it is reasonable to assume that this trend will continue. On the basis of United Nations estimates that the world's population will approximately double next century, we can estimate that, unless efforts are made to reduce human-related methane sources, the enhanced greenhouse effect of increased methane (which to date is about 20 per cent of that of increased carbon dioxide through human activity) will also approximately double over that time (Fig. 3.8).

NITROUS OXIDE

Nitrous oxide, used as a common anaesthetic and known as laughing gas, is another minor greenhouse gas. Its concentration in the atmosphere of about 0.3 ppmv is rising at about 0.25 per cent per year and is about 8 per cent greater than in pre-industrial times. It possesses a relatively long atmospheric lifetime of about 150 years. The additional sources which are leading to its increase are not well identified although the chemical industry (for example, nylon production), deforestation and agricultural practices all play some part. To estimate its likely increase next century we need first to know much more detail about the human-generated sources responsible for its production.

CHLOROFLUOROCARBONS (CFCs) AND OZONE

The CFCs are man-made chemicals which, because they vaporize just below room temperature and because they are non-toxic and non-flammable, are ideal for use in refrigerators and aerosol spray cans. Since they are so chemically unreactive, once they are released into the atmosphere they remain for a long time—one or two hundred years—before being destroyed. As their use has increased rapidly through the 1980s their concentration in the atmosphere has been building up so that they are now present (adding together all the different CFCs) in about 1ppbv (part per thousand million by volume). This may not sound very much, but it is quite enough to cause two serious environmental problems.

The first problem is that they destroy ozone. Ozone (O_3), a molecule consisting of three atoms of oxygen, is an extremely reactive gas present in small quantities in the stratosphere (a region of the atmosphere between about 10km and 50km in altitude). Ozone molecules are formed through the action of ultraviolet radiation from the sun on molecules of oxygen. They are in turn destroyed by a natural process as they absorb solar ultraviolet radiation at slightly longer wavelengths—radiation which would otherwise be harmful to us and to other forms of life at the Earth's surface. The amount of ozone in the stratosphere is determined by the balance between these two processes, one forming ozone and one destroying it. What happens when CFC molecules move into the stratosphere is that some of the chlorine atoms they contain are stripped off, also by the action of ultraviolet sunlight. These chlorine atoms readily react with ozone, reducing it back to oxygen and adding to the rate of destruction of ozone. This occurs in a catalytic cycle—one chlorine atom can destroy many molecules of ozone.

The ozone hole

The problem of ozone destruction was brought to world attention in 1985 when Joe Farman, Brian Gardiner and Jonathan Shanklin at the British Antarctic Survey discovered a region of the atmosphere over Antarctica where, during the southern spring, about half the ozone overhead disappeared. The existence of the 'ozone hole' was a great surprise to the scientists; it set off an intensive investigation into its causes. The chemistry and dynamics of its formation turned out to be complex. They have now been unravelled, at least as far as their main features are concerned, leaving no doubt that chlorine atoms introduced into the atmosphere by human activities are largely responsible. Not only is there depletion of ozone in the spring over Antarctica but also substantial reduction, of the order of 5 per cent, of the total column of ozone—the amount above one square metre at a given point on the Earth's surface—

at mid latitudes in both hemispheres.

Because of these serious consequences of the use of CFCs, international action has been taken. Many governments have now signed the Montreal Protocol set up in 1987 which, together with the Amendments agreed in London in 1991 and in Copenhagen in 1992, requires that manufacture of CFCs be phased out completely by the year 1996 in industrialized countries and by 2006 in developing countries. Because of this action the concentration of CFCs in the atmosphere should soon stop increasing. However, since they possess a long life in the atmosphere, little decrease will be seen for some time and substantial quantities will be present well over a hundred years from now.

So much for the problem of ozone destruction. The other problem with CFCs and ozone, the one which concerns us here, is that they are both greenhouse gases. They possess absorption bands in the region known as the atmospheric window (see Fig. 2.4) where few other gases absorb. Because, as we have seen, the CFCs destroy some ozone, the greenhouse effect of the CFCs is partially compensated by the reduced greenhouse effect of atmospheric ozone.

First considering the CFCs on their own, a CFC molecule added to the atmosphere has a greenhouse effect five to ten thousand times greater than an added molecule of carbon dioxide. Thus, despite their very small concentration compared, for instance, with carbon dioxide, they have a significant greenhouse effect. It is estimated that a warming of about 0.2°C on average (or about 20 per cent of the total amount of

warming by all greenhouse gases) will have occurred in the tropics—at higher latitudes there is a compensating effect due to ozone reduction, explained below—due to the levels of CFCs present in the atmosphere by the end of the 1990s. This warming will only decrease very slowly next century.

Turning now to ozone, the effect from ozone depletion is complex because the amount by which ozone greenhouse warming is reduced depends critically on the level in the atmosphere at which it is being destroyed. Further, ozone depletion is concentrated at high latitudes while the greenhouse effect of the CFCs is uniformly spread over the globe. In tropical regions there is virtually no ozone depletion so no change in the ozone greenhouse effect. At mid-latitudes, very approximately, the greenhouse effects of ozone reduction and of the CFCs compensate for each other. In polar regions the reduction in the greenhouse effect of ozone more than compensates for the greenhouse warming effect of the CFCs[17].

As CFCs are phased out, they are being replaced to some degree by other halocarbons—hydrochloro-fluorocarbons (HCFCs) and hydrofluorocarbons (HFCs). In Copenhagen in 1992, the international community decided that HCFCs would also be phased out by the year 2030. While being less destructive to ozone than the CFCs, they are still greenhouse gases. The HFCs contain no chlorine or bromine, so they do not destroy ozone and are not covered by the Montreal Protocol. Because of their shorter lifetime, typically tens rather than hundreds of years, the concentration in the atmosphere of both the HCFCs and the HFCs, and therefore their contribution to global warming for a given rate of emission, will be less than for the CFCs[17]. However, should their rate of production threaten to rise to much larger values, their potential contribution to greenhouse warming must be exposed and properly considered.

Ozone is also present in the lower atmosphere or troposphere, where some of it is transferred downwards from the stratosphere and where some is generated by chemical action, particularly as a result of the action of sunlight on the oxides of nitrogen. It is especially noticeable in polluted atmospheres; if present in high enough concentration, it can become a health hazard. There is some evidence that ozone near the Earth's surface has been increasing in recent decades, but we do not yet know if it is significant on a global scale from a greenhouse warming point of view. Higher up in the troposphere where aircraft have their cruising levels, nitrogen oxides are being emitted from aircraft and ozone is being created[18]. Estimates of the greenhouse effect of the additional ozone suggest that it may be large, comparable in fact to the greenhouse effect of the carbon dioxide emitted from the aircraft exhausts[19].

GASES WITH AN INDIRECT GREENHOUSE EFFECT

I have described all the gases present in the atmosphere which have a direct greenhouse effect. There are also gases which through their chemical action on greenhouse gases, for instance on methane or on lower atmospheric ozone, have an influence on the overall size of greenhouse warming. Carbon monoxide (CO) and the nitrogen oxides (NO and NO_2) emitted, for instance, by motor vehicles, are some of these. Carbon monoxide has no direct greenhouse effect of its own but, as a result of chemical reactions, it forms carbon dioxide. These reactions also affect the amount of the hydroxyl radical (OH) which in turn affects the concentration of methane.

Much research needs to be done on many of the chemical processes in the atmosphere in order to elucidate the details of these processes which have indirect effects on greenhouse gases. While it is important to do this and to take them properly into account, it is also important to recognize that their combined effect is much less than that of the major contributors to human-generated greenhouse warming, namely carbon dioxide and methane.

Particles in the atmosphere

Particles in the atmosphere affect its energy balance because they both absorb radiation from the sun and scatter it back to space. We can easily see the effect of this on a bright day in the summer with a light wind when downwind of an industrial area. Although no cloud appears to be present, the sun appears hazy. We call it 'industrial haze'. Under these conditions a significant proportion of the sunlight incident at the top of the atmosphere is being lost as it is scattered back and out of the atmosphere by the millions of small particles in the haze.

Atmospheric particles come from a variety of sources. They are blown off the land surface, especially in desert areas; they result from forest fires and they come from sea spray. From time to time large quantities of particles are injected into the upper atmosphere from volcanoes. Some particles are also formed in the atmosphere itself.

Sulphate particles are particularly important. They are formed as a result of chemical action on sulphur dioxide, a gas which is produced in large quantities by power stations and other industry in which coal and oil (both of which contain sulphur in varying quantities) is burnt. Sulphur dioxide also creates another pollution problem; it is the main source of acid rain which is leading to the degradation of forests and fish stocks in lakes downwind of major industrial areas.

The number of sulphate particles in the atmosphere is increasing due to the burning of fossil fuels which contain sulphur. Their effect is mainly confined to the major industrial regions of the northern hemisphere. Averaged over the northern hemisphere the amount of sunlight scattered back to space by the sulphate particles from human-generated sources has been estimated to amount to about 0.4 watt per square metre[20]. Over limited regions of the northern hemisphere, therefore, the radiative effect of these particles is comparable in size, although opposite in effect, to that of human-generated greenhouse gases up to the present time. Because of the large regional variation of the sulphate particles, any effect they have on the climate can also be expected to be strongly regional in character—studies of this with climate models need to be carried out.

Because of the problems of acid rain associated with sulphur dioxide emissions, serious efforts are

currently underway, especially in Europe and North America, to curb these emissons to a substantial degree. Although the amount of sulphur-rich coal being burnt elsewhere in the world is rising, the damaging effects of sulphur pollution are likely to lead to an extension of the controls on sulphur emissions. Although, therefore, these emissions may already have been responsible for some significant climate effect (see Chapter 6), further modification of the climate from this source is likely to be limited.

There is a further way by which particles in the atmosphere could influence the climate; through their effect on cloud formation[21]. If particles are present in large numbers when clouds are forming, the resultant cloud will consist of a large number of smaller drops than would otherwise be the case—this is similar to what happens as polluted fogs form in cities. Such a cloud will be more highly reflecting to sunlight than one consisting of larger particles, thus further increasing the energy loss resulting from the presence of the particles. It is not well known how large this effect is likely to be; it needs to be studied, especially by making careful measurements on suitable clouds.

These effects which may arise from particles in the atmosphere from human-generated sources may all have some influence on the climate—an influence which will be mainly on a local or regional rather than a global scale. Although they have not yet been included to any extent in projections of global climate change into the future, they must be taken into account in any detailed discussion of the impact of climate change in particular regions.

Radiative forcing from changes in greenhouse gas emissions

This chapter has summarized current scientific knowledge about the sources and sinks of the main greenhouse gases and the exchanges which occur between the components of the climate system—the atmosphere, the ocean and the land surface—including the close balances which are maintained between the different components and the way in which these balances are being disturbed by human-generated emissions. Different assumptions about future such emissions have been used to generate emission scenarios. From these scenarios estimates have been made (for carbon dioxide, for instance, using a computer model of the carbon cycle) of likely increases in greenhouse gas concentrations in the future.

Given information about the possible increases in greenhouse gases, the next step is to calculate the effect of these increases on the amounts of thermal (infrared) radiation absorbed and emitted by the atmosphere. This is done using information about how the different gases absorb radiation in the infrared part of the spectrum, as mentioned in Chapter 2. The overall effect due to each gas on the thermal radiation stream leaving the top of the atmosphere[22] is known as its *radiative forcing*. Fig. 3.8 shows the contributions to radiative forcing from different gases which would arise next century for the business-as-usual emission scenario IS 92a.

It is often convenient to make calculations of the greenhouse effect assuming that carbon dioxide is the only greenhouse gas. Because of this it is also useful to convert the other greenhouse gases to equivalent amounts of carbon dioxide, in other words to the amounts of carbon dioxide which would give the same radiative forcing. The information in Fig. 3.8 enables the conversion to be carried out[23]. For instance, the increases in all the greenhouse gases to date are equivalent to an increase in carbon dioxide of about 50 per cent and, under the IS 92a scenario, doubling of the equivalent carbon dioxide amount from pre-industrial times will occur in about 2030. Chapters 5 and 6 will explain how these estimates of radiative forcing can be incorporated into computer climate models to predict the future climate change which is likely to occur because of human activities. However, before considering predictions of future climate change, it is helpful to gain perspective by looking at some of the climate changes which have occurred in the past.

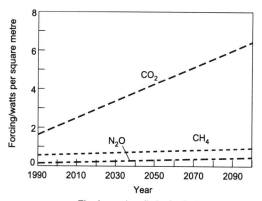

FIG. 3.8 The changes in radiative forcing in watts per square metre since pre-industrial times under the IPCC emissions scenario IS 92a for the greenhouse gases CO_2, CH_4 and N_2O[24]. In 1990, the amount of radiative forcing in the tropics due to the CFCs was about 0.22 watts per square metre.

FOOTNOTES

1 Adapted from S.H. Schneider 'The changing climate' *Sci. Amer.* September 1989 pp 38–47 and from U. Siegenthaler and J.L.Sarmiento, 'Atmospheric carbon dioxide and the ocean', *Nature*, **365**, 1993, pp 119–25.

2 From R.T. Watson *et al.* 'Greenhouse gases and aerosols' in *Climate Change, the IPCC Scientific Assessment*, ed. J.T. Houghton, G.J. Jenkins and J.J. Ephraums, CUP, 1990, pp 1–40 (updated).

3 The data for diagram (b) are from C.D. Keeling of the Scripps Institute of Oceanography and P. Tans of the National Oceanic and Atmospheric Administration of the United States.

4 J. Woods and W. Barkmann, 'The plankton multiplier—positive feedback in the greenhouse', *J. of Plankton Research,* **15**, 1993, pp 1053–74.

5 Adapted from U. Siegenthaler and J.L. Sarmiento, 'Atmospheric carbon dioxide and the ocean', *Nature* **365**, 1993, pp 119–25.

6 Adapted from U. Siegenthaler and J.L. Sarmiento *loc. cit.*.

7 For a review of the present knowledge of the carbon cycle see E.T. Sundquist, 'The global carbon dioxide budget', *Science*, **259**, 1993, pp 934–40.

8 For more information on biological feedbacks see *Biotic Feedbacks*, eds G.M. Woodwell and F.T. Mackenzie, OUP, 1994.

9 J. Leggett, W.J. Pepper & R.J. Swart, 'Emissions Scenarios for the IPCC: an update' in *Climate Change 1992, The Supplementary Report to the IPCC Scientific Assessment*, eds J.T. Houghton, B.A. Callander, S.K.Varney, pp 69–95, CUP, 1992.

10 World Energy Council Commission Report, *Energy for tomorrow's world*, World Energy Council, London, 1993.

11 For more information on the model see T.M.C. Wigley, 'Balancing the carbon budget: implications for projections of future carbon dioxide changes', *Tellus*, **45B**, 1993, pp 409–425.

12 From World Energy Council, *op cit*, 1993, Appendix E, pp 305–306. The model used in these calculations was developed by Professor T.Wigley.

13 From R.T. Watson *et al., op. cit.* 1990—updated.

14 R.T. Watson *et al.*, 'Greenhouse gases: sources and sinks' in *Climate Change 1992, The Supplementary Report to the IPCC Scientific Assessment*, eds J.T. Houghton, B.A. Callander, S.K. Varney, CUP, 1992, pp 25–46.

15 During the last year or so to 1993, the increase has slowed to almost zero. The reason for this is not known but one suggestion is that, because of recent changes in Russia, the leakage from Siberian natural gas pipelines has been much reduced.

16 The ratio of the enhanced greenhouse effect of a molecule of methane compared to a molecule of carbon dioxide is known as its global warming potential (GWP). The figure of about 7.5 given here for the GWP of methane is for a time horizon of 100 years—reference J. Lelieveld and P.J. Crutzen, *Nature*, **355**, 1992, pp 339–41. About half of the contribution of methane to the greenhouse effect is because of its direct effect on the outgoing thermal radiation. The other half arises because of its influence on the overall chemistry of the atmosphere. Increased methane eventually results in small increases in water vapour in the upper atmosphere, in tropospheric ozone and in carbon dioxide, all of which in turn add to the greenhouse effect.

17 More detail on this and the radiative effects of minor gases and particles can be found in I.S.A. Isaksen *et al.*, 'Radiative Forcing of Climate' in *Climate Change 1992, The Supplementary Report to the IPCC Scientific Assessment*, eds J.T. Houghton, B.A. Callander, S.K. Varney, CUP, 1992, pp 47–67.

18 W-C. Wang and Y-C. Zhuang, *Geophysical Research Letters* **20**, 1993, pp 1567–70.

19 C.E. Johnson, J.E. Henshaw and G. McInnes, *Nature*, **355**, 1993, pp 69–71.

20 J.T. Kiehl and B.P. Briegleb, 'The relative roles of sulphate aerosols and greenhouse gases in climate forcing', *Science*, **260**, 1993, pp 311–14. Earlier work by R.Charlson quoted in I.S.A.Isaksen *et al., loc. cit.* estimated the radiative effects of anthropogenic sulphate aerosol to be about twice those of Kiehl and Briegleb.

21 More details in K.P. Shine *et al.*, 'Radiative forcing of climate' in *Climate Change, the IPCC Assessment, op. cit.*, 1990, pp 41–68.

22 Strictly it is the net effect at the top of the troposphere which is usually calculated.

23 The assumption that greenhouse gases may be treated as equivalent to each other is a good one for many purposes. However, because of the differences in their radiative properties, accurate modelling of their effect should treat them separately. More details of this problem in W.L.Gates *et al.*, 'Climate modelling, climate prediction and model validation' in *Climate change 1992, The Supplementary Report to the IPCC Scientific Assessment*, eds J.T. Houghton, B.A. Callander, S.K. Varney, CUP, 1992, pp 171–75.

24 J.F.B. Mitchell and J.M. Gregory, 'Climatic consequences of emissions', in *Climate Change 1992, The Supplementary Report to the IPCC Scientific Assessment*, eds J.T. Houghton, B.A. Callander, S.K. Varney, CUP, 1992, pp 171–75. These calculations of forcing are based on estimates of carbon dioxide concentrations (given in the same article in the *IPCC 1992 Supplementary Assessment*) which are higher than those for IS92a shown in Fig. 3.6; they are close to the high end of the range of uncertainty shown in Fig. 3.6.

4 Climates of the Past

To obtain some perspective against which to view future climate change, it is helpful to look at some of the climate changes which have occurred in the past. This chapter will briefly consider climatic records and climate changes in three periods: the last hundred years, then the last thousand years and finally the last million years. At the end of the chapter some interesting recent evidence for the existence of relatively rapid climate change at various times during the past one or two hundred thousand years will be presented.

The last hundred years

The 1980s and early 1990s have brought some unusually warm years for the globe as a whole (see Chapter 1); seven of the eight warmest years during the past century have occurred during this period (up to and including 1992). 'Warmest' means only a few tenths of a degree Celsius at most, but in terms of the global average such differences are quite significant. Fig. 4.1 shows the global average temperature since 1860. An increase over this period has taken place of between 0.3°C and 0.6°C. Although there is a distinct trend, the increase is by no means a uniform one; in fact, some periods of cooling as well as warming have occurred.

A sceptic may wonder how a diagram like Fig. 4.1 can be prepared and whether any reliance can be placed upon it. After all, temperature varies from place to place, from season to season and from day to day by many tens of degrees. How, then, can a global average changing by a few tenths of a degree have any meaning?

First of all, just how is a change in global average temperature estimated from a combination of records of changes in the near surface temperature over land and changes in the temperature of the sea surface (Fig. 4.1)? To estimate the changes over land, weather stations are chosen where consistent observations have been taken from the same location over a substantial proportion of the whole 130 year period. Changes in sea surface temperature have been estimated by processing over sixty million observations from ships—mostly merchant ships—over the same period. All the observations, from land stations and from ships, are then located within a grid of squares, say 1° of latitude by 1° of longitude, covering the Earth's surface. Observations within each square are averaged; the global average is obtained by averaging (after weighting them by area) over the averages for each of the squares.

A number of research groups in the world in different countries have made careful and independent analyses of these observations. In somewhat different ways they have made allowances for factors which could have introduced artificial changes in the records. For instance, the record at some land stations could have been affected by changes in their surroundings as these have become more urban. In the case of ships, the standard method of observation used to be to insert a thermometer into a bucket of water taken from the sea. Small changes of temperature have been shown to occur during this process; the size of the changes varies between day and night and is also dependent on several other factors including the material from which the bucket is made. Over the years wooden, canvas and metal buckets have been variously employed; nowadays, a large proportion of the observations are made by measuring the temperature of the water entering the engine cooling system. Careful analysis of the effects of these details on observations both on land and from ships has enabled appropriate corrections to be made to the record, and good agreement has been achieved between analyses carried out at different centres.

Confidence that the observed variations are real is increased by noticing that the trend and the shape of the changes are similar when different selections of the total observations are made. For instance, the separate records from the land and sea surface (Fig. 4.1), and from the northern and southern hemispheres are closely in accord.

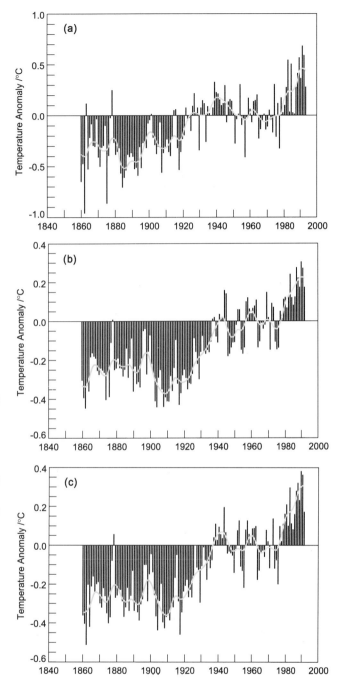

FIG. 4.1 Changes in global average temperature from 1860 to 1992 relative to the period 1951–80. (a) over land (b) for the sea-surface (c) land and sea temperatures combined[1].

During the last twenty years or so observations have been available from satellites orbiting around the Earth. Their great advantage is that they automatically provide data with global coverage, which is often lacking in other data sets. The length of the record from satellites, however, is generally less than twenty years, a comparatively short period in climate terms. It has been suggested that satellite measurements of lower atmospheric temperature since 1979 are not consistent with the trend of rising temperatures in surface observations; a careful analysis, however, does not confirm this as a significant inconsistency (see box).

The most obvious feature of the climate record illustrated in Fig. 4.1 is that of considerable variability, not just from year to year, but from decade to decade. Some of this variability will have arisen through causes external to the atmosphere and the oceans, for instance, as a result of volcanic eruptions such as those of Krakatoa in 1883 or of Pinatubo in the Philippines in 1991 (the low global average temperature in 1992, compared with 1990 and

1991, is very likely due to the Pinatubo volcano). But there is no need to invoke volcanoes or other external causes to explain most of the variations in the record. Their most likely cause is internal variations within the total climate system, for instance between different parts of the ocean[2].

Because of the degree of natural variability shown by the climate record, it is clearly difficult to be certain that the increasing trends in global average temperature over the past century and over the past decade are due to global warming arising from the increase of greenhouse gases. For instance, the particular increase from 1910 to 1940 (Fig. 4.1) is too rapid to have been due to the rather small increase of greenhouse gases during that period. Eventually, as the greenhouse gases increase further, the amount of warming is expected to become sufficiently large that it will swamp

Atmospheric temperature observed by satellites

Since 1979 meteorological satellites flown by the National Oceanic and Atmospheric Administration (NOAA) of the United States have carried a microwave instrument, the Microwave Sounding Unit (MSU), for the remote observation of the average temperature of the lower part of the atmosphere. Fig. 4.2 shows the record of global average temperature deduced from the MSU and compares it with data from balloon sondes and surface data for the same period. Considering the possible errors in the measurements and also considering that the three measurements do not cover exactly the same part of the atmosphere, good agreement is shown between the plots from the three very different sources. The plots also illustrate the impossibility of deducing information about trends from such records when they cover only just over ten years and when they show substantial variability.

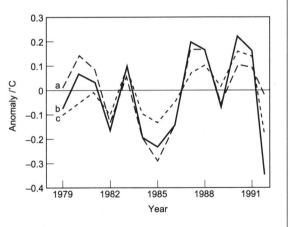

FIG. 4.2 Comparison, for 1979–1992, of global average temperature (shown as deviations from the 1979–1989 average) from (a) the MSU channel covering the lower atmosphere, (b) balloon borne radiosondes, averaged over the lower atmosphere, (c) surface data as Fig. 4.1[3].

the natural variations in climate; then it will be clear that global warming has almost certainly been observed. In the meantime, the global average temperature may continue to increase or it could show periods of decrease. However, over the next few years and the next decade or two scientists will be inspecting climate changes and climate events most carefully as they occur, to try to detect signs of global warming and to see to what extent actual events can be related to scientific predictions. The details of these predictions will be discussed in later chapters.

The last thousand years

The detailed systematic record of weather elements such as temperature, rainfall, cloudiness and the like presented above for the last hundred years and which covers a good proportion of the globe is not available for earlier periods. Further back, the record becomes much more sparse and doubt arises over the consistency of the instruments used for observation. Most thermometers in use two hundred years ago were not well calibrated or carefully exposed. However, many diarists and writers kept records at different times; from a wide variety of sources weather and climate information can be pieced together. Indirect sources, such as are provided by ice cores, tree rings and records of lake levels, of glacier advance and retreat or of pollen distribution in the past, can also yield information to assist in building up the whole climatic story. From a variety of sources, for instance, it has been possible to put together a systematic atlas of weather patterns covering the last five hundred years for China.

Similarly, from direct and

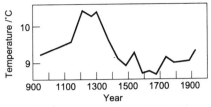

FIG. 4.3 Average temperature in Central England during the last thousand years[4].

indirect sources, Professor Hubert Lamb of the Climatic Research Unit of the University of East Anglia has derived a record of the average temperatures for Central England for the past thousand years (Fig. 4.3). Its main features are of a medieval warm period between about AD1100 and 1300 when vines were grown as far north as Yorkshire, and the 'Little Ice Age' between about AD1400 and 1850, during which freezing of the Thames in winter was not uncommon. There is plenty of evidence that the 'Little Ice Age' extended throughout Europe and North America; indications from glaciers in other parts of the world including the southern hemisphere suggest that it could have been experienced worldwide.

There is as yet no certain explanation for these warm and cold periods during the past thousand years. Volcanic activity has sometimes been suggested as a possible cause. One of the largest eruptions during the period was that of Tambora in Indonesia in April 1815, which was followed in many places by two exceptionally cold years; 1816 was described in New England and Canada as the 'year without a summer'. But the effect on the climate even of an eruption of the magnitude of Tambora only lasts a few years at most. Others have postulated that variations in the output of energy from the sun might

FIG. 4.4 **Observations from the Vostok ice core,** showing for the last 160,000 years the variation of atmospheric temperature over Antarctica (it is estimated that the variation of global average temperature would be of the order of half that in the polar regions), and the atmospheric carbon dioxide and methane concentrations[5]. The thickness of the lines for the carbon dioxide and methane plots indicates the range of uncertainty in the measurements.

be a cause. However, there is no direct evidence for such variations so their possibility must remain speculative. Nor can it be suggested that greenhouse gases such as carbon dioxide and methane have been the cause of change; for the millennium before 1800 their concentration in the atmosphere was rather stable, the carbon dioxide concentration, for instance, varying by less than 3 per cent. The most likely cause is, in fact, not due to any factors external to the atmosphere. As with the shorter-term changes mentioned earlier, such variations of climate can arise naturally from internal variations

within the atmosphere and the ocean and in the two-way relationship—coupling—between them.

The past million years

To go back before recorded human history, scientists have to rely on indirect methods to unravel much of the story of the past climate. A particularly valuable information source is the record stored in the ice which caps Greenland and the Antarctic continent. These icecaps are several thousands of metres thick. Snow deposited on their surface gradually becomes compacted as further snow falls, becoming solid ice. The ice moves steadily downwards, eventually flowing outwards at the bottom of the ice-sheet. Ice near the top of the layer will have been deposited fairly recently; ice near the bottom will have fallen on the surface many tens of thousands of years ago. Analysis of the ice at different levels can, therefore, provide information about the conditions prevailing at different times in the past.

Deep cores have been drilled out of the ice at several locations in both Greenland and Antarctica. At Russia's Vostok station in east Antarctica, for instance, drilling has been carried out for about twenty years. The longest and most recent core reached a depth of over 2.5km; the ice at the bottom of the hole fell as snow on the surface of the Antarctic continent over 200,000 years ago. More recently cores of similar length from the Greenland icecap have been drilled by European and American teams.

Small bubbles of air are trapped within the ice. Analysis of the composition of that air shows what was present in the atmosphere for the time at which the ice was formed—gases such as carbon dioxide or

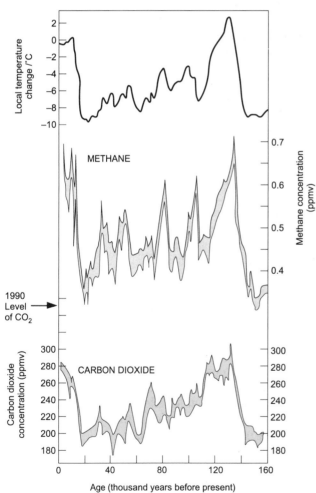

methane, or dust particles that may have come from volcanoes or from the sea surface. Further information is provided by analysis of the ice itself. Small quantities of different oxygen isotopes and of the heavy isotope of hydrogen (deuterium) are contained in the ice. The ratios of these isotopes that are present depend sensitively on the temperatures at which evaporation and condensation took place for the water in the clouds from which the ice originated (see box). These in turn are dependent on the average temperature near the surface of the Earth. A temperature record for the polar regions can therefore be constructed from analyses of the ice cores.

Such a reconstruction from a Vostok core for the temperature and the content of carbon dioxide and methane is shown in Fig. 4.4 for the past 160,000 years, which includes the period of the last climatic cycle. It shows the last major ice age which began about 120,000 years ago and began to come to an end about 20,000 years ago. It also demonstrates the close connections which exist between temperature and carbon dioxide and methane concentrations.

Data from ice cores can take us back 200,000 years or so. To go further back, over the past million years, the composition of ocean sediments can be investigated to yield information. Fossils of plankton and other small sea creatures

Paleoclimate reconstruction from isotope data

The isotope ^{18}O is normally present in normal oxygen at a concentration of about 1 part in 500 compared with the more abundant isotope ^{16}O. When water evaporates, water containing the lighter isotope is more easily vaporized, so that water vapour in the atmosphere contains less ^{18}O than sea water. Similar separation occurs in the process of condensation when ice crystals form in clouds. The amount of separation between the two oxygen isotopes in these processes depends on the temperature at which evaporation and condensation occur. Measurements on snowfall in different places can be used to calibrate the method; it is found that the concentration of ^{18}O varies by about 0.7 of a part per thousand for each degree of change in average temperature at the surface. Information is therefore available in the ice cores taken from polar ice-caps concerning the variation in atmospheric temperature in polar regions during the whole period when the ice core was laid down.

Since the ice-caps are formed from accumulated snowfall which contains less ^{18}O than sea water, the concentration of ^{18}O in water from the oceans provides a measure of the total volume of the ice in the ice-caps; it changes by about one part in 1,000 between the maximum ice extent of the ice ages and the warm periods in between. Information about the ^{18}O content of ocean water at different times is locked up in corals and in cores of sediment taken from the ocean bottom, which contain carbonates from fossils of plankton and small sea creatures from past centuries and millennia. Measurements of radioactive isotopes, such as the carbon isotope ^{14}C, and correlations with other significant past events enable the corals and sediment cores to be dated. Since the separation between the oxygen isotopes which occurs as these creatures are formed also depends on the temperature of the sea water (although the dependence is weaker than the other dependencies considered above) information is also available about the distribution of ocean surface temperature at different times in the past[6].

deposited in these sediments also contain different isotopes of oxygen. In particular the amount of the heavier isotope of oxygen (^{18}O) compared with the more abundant isotope (^{16}O) is sensitive both to the temperature at which the fossils were formed and to the total volume of ice in the world's ice-caps at the time of the fossils' formation (see box).

From the variety of paleoclimate data which is available, variations in the volume of ice in the ice-caps can be reconstructed over the greater part of the last million years (Fig. 4.5c). In this record six or seven major ice ages can be identified with warmer periods in between, the period between these major ice ages being

approximately 100,000 years. Other cycles are also evident in the record.

The most obvious place to look for the cause of regular cycles in climate is outside the Earth, in the sun's radiation. Has this varied in the past in a cyclic way? So far as is known the output of the sun itself has not changed to any significant extent over the last million years or so. But because of variations in the Earth's orbit, the distribution of solar radiation has varied in a more or less regular way during the last millennium.

Three regular variations occur in the orbit of the Earth around the sun (Fig. 4.5a). The Earth's orbit, although nearly circular, is actually an ellipse. The eccentricity of the ellipse (which is related to the ratio between the greatest and the least diameters) varies with a period of about 100,000 years; that is the slowest of the three variations. The Earth also spins on its own axis, the axis of spin being tilted with respect to the axis of the Earth's orbit, the angle of tilt varying between 21.6° and 24.5° (currently it is 23.5°) with a period of about 41,000 years. The third variation is of the time of year when the Earth is closest to the sun (the Earth's perihelion). In the present configuration, the Earth is closest to the sun in January; the time of perihelion moves through the months of the year with a period of about 23,000 years (see also Fig. 5.19).

As the Earth's orbit changes its relationship to the sun, although the total quantity of solar radiation reaching the Earth varies very little, the distribution of that radiation with latitude and season over the Earth's surface changes considerably. The changes are especially large in polar regions where the variations in summer sunshine, for instance, reach

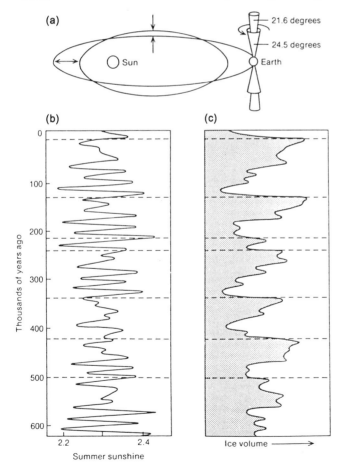

(a)

21.6 degrees
24.5 degrees
Sun
Earth

(b)

Thousands of years ago

2.2   2.4
Summer sunshine

(c)

Ice volume ⟶

about 10 per cent (Fig. 4.5b). James Croll, a British scientist, first pointed out in 1867 that the major ice ages of the past might be linked with these regular variations in the seasonal distribution of solar radiation reaching the Earth. His ideas were developed in 1920 by Milankovitch, a climatologist from Yugoslavia, whose name is usually linked with the theory. Inspection by eye of the relationship between the variations of polar summer sunshine and global ice volume shown in Fig. 4.5 indicates a strong connection. Careful study of the correlation between the two curves confirms this and demonstrates that 60 per cent of the variance in the climatic record of global ice volume falls close to the three frequencies of regular variations in the Earth's orbit, thus providing support for the Milankovitch theory.

More careful study of the relationship between the ice ages and the Earth's orbital variations shows that the size of the climate changes is larger than might be expected from forcing by the radiation changes alone. Other processes which could enhance the effect of the radiation changes (in other words, positive feedback processes) have to be introduced to explain the climate variations. The strong correlation observed in the climatic record between average atmospheric temperature and carbon dioxide concentration (Fig. 4.4) suggests that one such feedback arises from the changes in carbon dioxide influencing atmospheric temperature through the greenhouse effect. Such a correlation does not prove the existence of the greenhouse feedback; in fact part of the correlation arises because the atmospheric carbon dioxide concentration is itself

influenced, through biological feedbacks (see Chapter 3), by factors related to the average global temperature. However, as we shall see in later chapters, climates of the past cannot be modelled successfully without taking into account the influence on climate of variation in carbon dioxide concentration[8].

How stable has past climate been?

The major climate changes considered so far in this chapter have taken place relatively slowly. The growth and recession of the large polar ice-sheets between the ice ages and the intervening warmer interglacial periods have taken on average many thousands of years. Over the last 8000 years or so, which can be investigated more thoroughly than earlier epochs, the climate has shown some slow changes but no dramatic fluctuations.

However, there is substantial evidence from data from the Greenland ice cores that the last 8000 years have been unusually stable and that much larger variations have been typical of earlier epochs. At the summit of the Greenland icecap, the rate of accumulation of snow has been higher than that at the drilling locations in Antarctica. Because of this, a period of, say, ten years in the past is represented by a longer length of ice core from Greenland and more detail of variations over relatively short periods of a few years is therefore available. Further, it appears that the fluctuations of temperature over Greenland have been larger than those over Antarctica.

Variations of a few degrees Celsius taking place over a few hundred years during the ice age period about 30,000 years ago were reported from Greenland ice core data in 1989. But in 1993, Professor Dansgaard from Denmark and his colleagues found evidence that rapid variations of the Arctic temperature frequently occurred during the glacial period up to 100,000 years ago. These variations have been of up to five or six degrees Celsius and have sometimes occurred over periods of less than a hundred years (Fig. 4.6). Comparison between the results from ice cores drilled at different locations within the Greenland icecap confirm the details up to about 100,000 years ago. But the records from before that time may have been distorted by ice movements and it is not yet clear from the ice core record whether the periods of rapid temperature variation extend into the interglacial warm period around 120,000 years ago[11].

Another particularly interesting period of climatic history, more recently, is the Younger Dryas event (so called because it was marked by the spread of an arctic flower, the *Dryas octopetala*) which occurred over a period of about 1500 years between about 12,000 and 10,700 years ago. For 6,000 years before the start of this event the Earth had been warming up after the end of the last ice age. But then during the Younger Dryas period, as demonstrated from many different sources of paleoclimatic data, the climate swung back again into much colder conditions similar to those at the end of the last ice age (Fig. 4.7). The ice core record shows that at the end of the event, 10,700 years ago, the warming in the Arctic of about 7°C occurred over only about 50 years and was associated with decreased storminess (shown by a dramatic fall in the amount of dust in the ice core) and an increase of precipitation of about 50 per cent.

Two main possible reasons for these rapid variations in the past have been suggested. One suggestion particularly applicable to ice age conditions is that, as the ice-sheets over Greenland and eastern Canada have built up, major break-ups have occurred from time to time, releasing massive numbers of icebergs into the north Atlantic. The second possibility is that the ocean circulation in the north Atlantic region has been strongly affected by injections of fresh water from the melting of ice. At present the ocean circulation here is strongly influenced by cold salty water sinking to deep ocean levels because its saltiness makes it dense; this sinking process is part of the 'conveyor belt' which is the major feature of the circulation of deep ocean water around the world (see Fig. 5.18). Large quantities of fresh

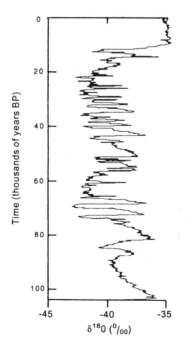

FIG. 4.6 Variations in Arctic temperature over the past 100,000 years as deduced from oxygen isotope measurements (in terms of $\delta^{18}O$)[9] from the 'Summit' ice core from Greenland. The overall shape of the record is similar to that from the Vostok ice core shown in Fig. 4.4 but much more detail is apparent in the 'Summit' record's stable period over the last 8,000 years[10]. A change of 5 parts per thousand in $\delta^{18}O$ in the ice core corresponds to about a 7°C change in temperature.

Time (thousands of years BP)

0

20

40

60

80

100

-45   -40   -35

$\delta^{18}O$ (°/oo)

water from the melting of ice would make the water less salty, preventing it from sinking and thereby altering the whole Atlantic circulation.

This link between the melting of ice and the ocean circulation is a key feature of the explanation put forward by Professor Wallace Broecker for the Younger Dryas event[12]. As the great ice-sheet over north America began to melt at the end of the last ice age, the melt water at first drained through the Mississippi into the Gulf of Mexico. Eventually, however, the retreat of the ice opened up a channel for the water in the region of the St Lawrence river. This influx of fresh water into the north Atlantic reduced its saltiness, thus, Broecker postulates, cutting off the formation of deep water and that part of the ocean 'conveyor belt'[13]. Warm water was therefore prevented from flowing northward, resulting in a reversal to much colder conditions. The suggestion is also that a reversal of this process with the starting up of

the Atlantic 'conveyor belt' could lead to a sudden onset of warmer conditions.

Although debate continues regarding the details of the Younger Dryas event, there is considerable evidence from paleodata, especially those from ocean sediments, for the main element of the Broecker explanation which involves the deep ocean circulation. It is also clear from paleodata that large changes have occurred at different times in the past in the formation of deep water and in the deep ocean circulation. Chapter 3 mentioned the possibility of such changes being induced by global warming through the growth of greenhouse gas concentrations. Our perspective regarding the possibilities of future climate change needs to take into account the rapid climate changes which have occurred in the past.

Having now in these early chapters set the scene, by describing the basic science of global warming, the greenhouse gases and their

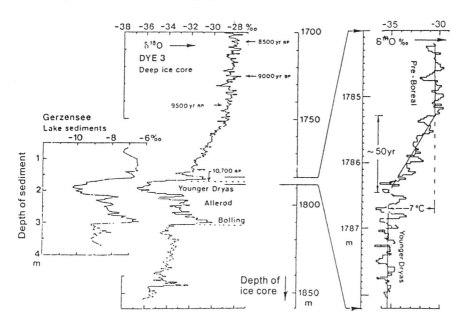

FIG. 4.7 Records of the variations of the oxygen isotope [18]O from lake sediments from Lake Gerzen in Switzerland and from the Greenland ice core 'Dye 3' showing the 'Younger Dryas' event and its rapid end about 10,700 years ago. Dating of the ice core was by counting the annual layers down from the surface; dating of the lake sediment was by the ^{14}C method. A change of 5 parts in a thousand in $\delta^{18}O$ in the ice core corresponds to about a 7°C change in temperature[14].

origins and the current state of knowledge regarding past climates, I move on in the next chapter to describe how, through computer models of the climate, predictions can be made about what climate change can be expected in the future.

FOOTNOTES

1 From C.K. Folland, Hadley Centre, U.K. Meteorological Office.

2 More details of this in Chapter 5.

3 From C.K. Folland et al. 'Observed climate variability and change' in *Climate Change 1992, the Supplementary Report to the IPCC Scientific Assessment*, eds J.T. Houghton, B.A. Callander and S.K. Varney, CUP, 1992; pp 135–70 (updated to include 1992 by C.K. Folland).

4 From H.H. Lamb, 'The early Medieval warm epoch and its sequel', *Palaeogeography, Palaeoclimatology, Palaeoecology*, 1, 1965, pp 13–37.

5 From D. Raynaud et al., 'The ice core record of greenhouse gases', *Science* 259, 1993, pp 926–34.

6 For more information about paleoclimates and how they are investigated see, for instance, T.L. Crowley and G.R. North, *Paleoclimatology*, OUP, 1991.

7 Adapted from W.S. Broecker and G.H. Denton, 'What drives glacial cycles?' *Sci. Amer.* 262 1990, pp 43–50.

8 See also D. Raynaud et al. 1993, *loc. cit.*

9 The quantity $\delta^{18}O$ plotted in the diagrams 4.6 and 4.7 is the difference (in parts per thousand) between the $^{18}O/^{16}O$ ratio in the sample and the same ratio in a laboratory standard.

10 Adapted from Professor Dansgaard and colleagues, Greenland ice core (GRIP) members, 'Climate instability during the last interglacial period recorded in the GRIP ice core', *Nature*, **364**, 1993, pp 203–207.

11 P.M. Grootes et al., 'Comparison of oxygen isotope records from the GISP2 and GRIP Greenland ice cores', *Nature*, **366**, pp 552–54, 1993.

12 W.S. Broecker & G.H. Denton, 1990, *loc. cit.*

13 More information in Chapter 5, see especially Fig. 5.18.

14 Adapted from W. Dansgaard, J.W.C. White and S.J. Johnsen, 'The abrupt termination of the Younger Dryas climate event', *Nature*, **339**, 1989, pp 532–33.

5 *Modelling the Climate*

Chapter 2 *looked at the greenhouse effect in terms of a simple radiation balance. That gave an estimate of the rise in the average temperature at the surface of the Earth as greenhouse gases increase. But any change in climate will not be distributed uniformly everywhere; the climate system is much more complicated than that. More detail in climate change prediction requires very much more elaborate calculations using computers. The problem is so vast that the fastest and largest computers available are needed. But before computers can be set to work on the calculation, a model of the climate must be set up for them to use[1]. A model of the weather as used for weather forecasting will be used to explain what is meant by a numerical model on a computer, followed by the increase in elaboration required to include all parts of the climate system in the model.*

Modelling the weather

An English mathematician, Lewis Fry Richardson, set up the first numerical model of the weather. During his spare moments while working for the Friends' Ambulance Unit (he was a Quaker) in France during the First World War he carried out the first numerical weather forecast. With much painstaking calculation with his slide-rule, he solved the appropriate equations and produced a six-hour forecast. It took him six months— and then it was not a very good result. But his basic methods, described in a book published in 1922[2], were correct. To apply his methods to real forecasts, Richardson imagined the possibility of a very large concert hall filled with people, each person carrying out part of the calculation, so that the integration of the numerical model could keep up with the weather. But he was many years before his time! It was not until some forty years later that, essentially using Richardson's methods, the first operational weather forecast was produced on an electronic computer. Computers up to 100,000 times faster than the one used for that first forecast (Fig. 5.1) now run numerical models which are the basis of all weather forecasts.

Setting up a model of the atmosphere for a weather forecast (see Fig. 5.2), requires a mathematical description of the way in which

FIG. 5.1 Computers used by the UK Meteorological Office for numeral weather prediction research, since 1965 for operational weather forecasting and most recently for research into climate prediction.

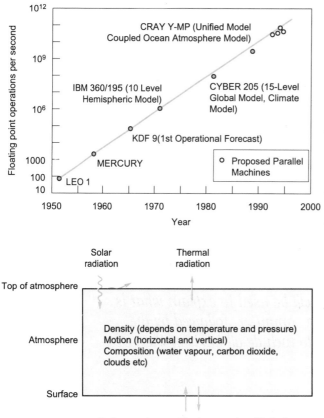

energy from the sun enters the atmosphere from above, some being reflected by the surface or by clouds and some being absorbed at the surface or in the atmosphere (see Fig. 3.6). The exchange of energy and water vapour between the atmosphere and the surface must also be described.

Water vapour is important because of its associated latent heat (in other words, it gives out heat when it condenses) and also because the condensation of water vapour results in cloud formation which modifies substantially the interaction of the atmosphere with the incoming energy from the sun. Variations in both these energy inputs modify the atmospheric temperature structure, causing changes in atmospheric density (since warmed gases expand and are therefore less dense). It is these density changes which drive atmospheric motions such as winds and air currents, which in their turn alter and feed back on atmospheric density and composition. More details of the model formulation are given in the box.

FIG. 5.2 Schematic illustrating the parameters and physical processes involved in atmospheric models.

Setting up a numerical atmospheric model

A numerical model of the atmosphere contains descriptions, in appropriate computer form and with necessary approximations, of the basic dynamics and physics of the different components of the atmosphere and their interactions[3]. When a physical process is described in terms of an algorithm (a process of step-by-step calculation) and simple parameters (the quantities which are included in a mathematical equation), the process is said to have been parametrized.

The dynamical equations are:

■ The horizontal momentum equations (Newton's second law of motion). In these,

the horizontal acceleration of a volume of air is balanced by the horizontal pressure gradient and the friction. Because the Earth is rotating, this acceleration includes the Coriolis acceleration. The 'friction' in the model mainly arises from motions smaller than the grid spacing, which have to be parametrized.

■ The hydrostatic equation. The pressure at a point is given by the mass of the atmosphere above that point. Vertical accelerations are neglected.

■ The continuity equation. This ensures conservation of mass.

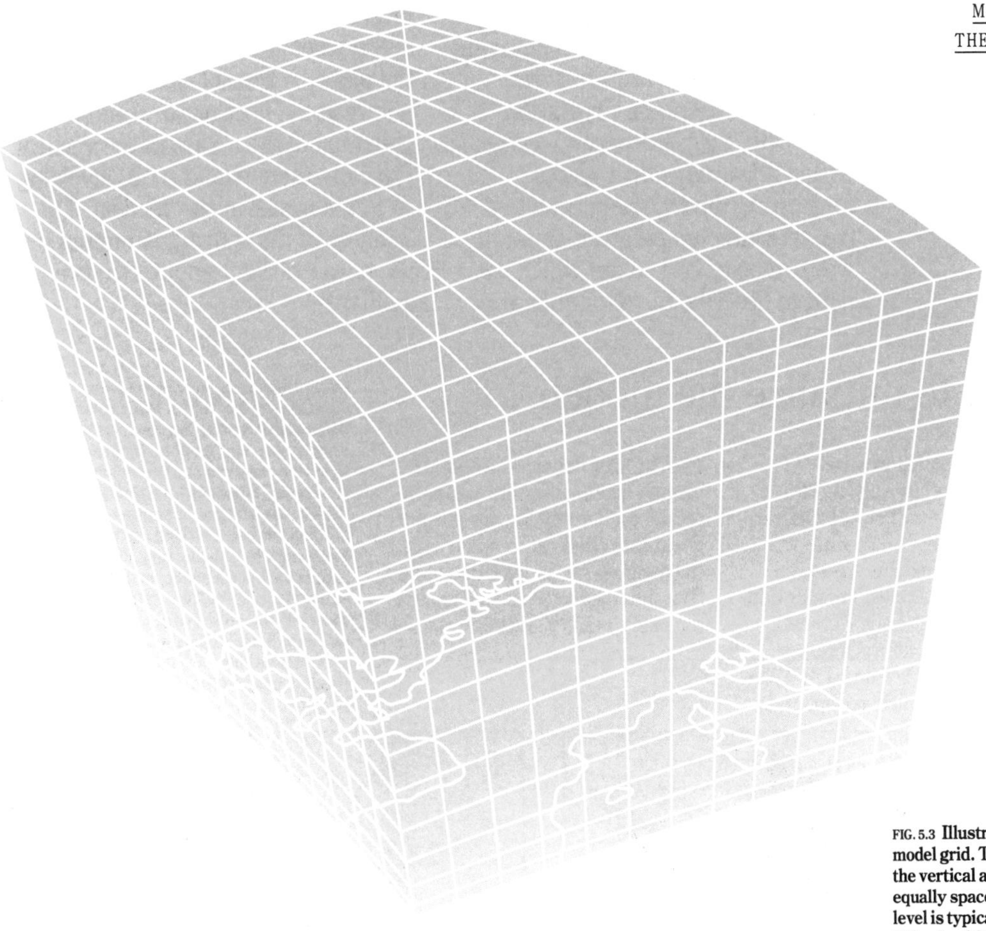

FIG. 5.3 **Illustration of a model grid. The levels in the vertical are not equally spaced; the top level is typically at about 30km in altitude.**

The model physics consists of:

■ The equation of state. This connects the quantities of pressure, volume and temperature for the gas of the atmosphere.

■ The thermodynamic equation (the law of conservation of energy).

■ Parametrization of moist processes (such as evaporation, condensation, formation and dispersal of clouds).

■ Parametrization of absorption and emission of solar radiation and of thermal radiation.

■ Parametrization of convective processes.

■ Parametrization of exchange of momentum (in other words, friction), heat and water vapour at the atmosphere/water surface.

Most of the equations in the model are differential equations, which means they describe the way in which quantities like pressure and wind velocity change with time and with location. If the rate of change of a quantity such as wind velocity and its value at a given time are known, then its value at a later time can be calculated. Constant repetition of this procedure is called integration. Integration of the equations is the process whereby new values of all necessary quantities are calculated at later times, providing the model's predictive powers.

To forecast the weather for several days ahead a model covering the whole globe is required; the southern hemisphere circulation today will affect European weather within a few days. In a global forecasting model, the parameters (i.e. pressure, temperature, humidity, wind velocity and so on) which we need to describe the dynamics and physics listed (see box) are specified at a grid of points (Fig. 5.3) covering the globe. A typical spacing between points in the horizontal would be 100km and about 1km in the vertical; typically there would be 20 levels or so in the model in the vertical. The fineness of the spacing is limited by the power of the computers currently available.

Data to initialize the model

At a major global weather forecasting centre such as the Meteorological Office at Bracknell, data from many sources are collected and fed into the model. This process is called initialization. Fig. 5.4 illustrates some of the sources of data for the forecast beginning at 12 UT on 1 July 1990. To ensure the timely receipt of data from around the world a dedicated communication network has been set up, used solely for this purpose. Great care needs to be taken with the methods for assimilation of the data into the model as well as with the data's quality and accuracy.

FIG. 5.4 **Illustrating some of the sources of data for input into the UK Meteorological Office global weather forecasting model on a typical day. Surface observations are from land observing stations (manned and unmanned), from ships and from buoys. Radiosonde balloons make observations up to 30km altitude from land and from ship-borne stations. Satellite soundings are of temperature and humidity at different atmospheric levels deduced from observations of infrared or microwave radiation. Satellite cloud-track winds are derived from observing the motion of clouds in images from geostationary satellites.**

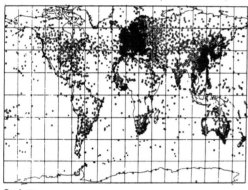

Surface

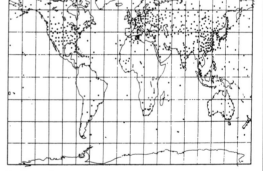

Radiosonde

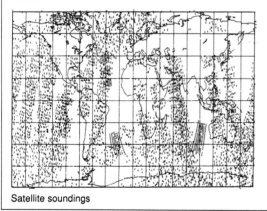

Satellite soundings

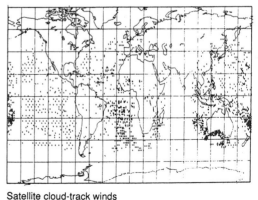

Satellite cloud-track winds

Now the model is set up, to generate a forecast from the present, it is started off from the atmosphere's current state and then the equations are integrated forward in time (see box) to provide new descriptions of the atmospheric circulation and structure up to six or more days ahead. For a description of the atmosphere's current state, data from a wide variety of sources (see box) have to be brought together and fed into the model.

Since computer models for weather forecasting were first introduced, their forecast skill has improved to an extent beyond any envisaged by those involved in the development of the early models. As improvements have been made in the model formulation, in the accuracy or coverage of the data used for initialization (see box) or in the resolution of the model (the distance between grid points) the resulting forecast skill has increased. For instance, for the British Isles, three-day forecasts of surface pressure today are as skilful on average as two-day forecasts of five years ago, as can be seen from Fig. 5.5.

When looking at the continued forecast improvement shown in Fig. 5.5, the question obviously arises as to whether the improvement will continue or whether there is a limit to the predictability we can expect. Because the atmosphere is a partially chaotic system (see box), even if perfect observations of the atmospheric state and circulation could be provided, there would be a limit to our ability to forecast the detailed state of the atmosphere at some time in the future. In Fig. 5.6 current forecast skill is compared with the best estimate of the limit of the forecast skill for the British Isles with a perfect model and near perfect data. According to that

estimate, the limit of significant future skill is about 20 days ahead.

Seasonal Forecasting

So far short-term forecasts of detailed weather have been under consideration. After 20 days or so they run out of skill. What about further into the future? Although we cannot expect to

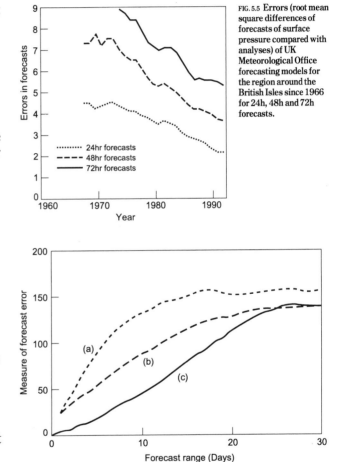

FIG. 5.5 Errors (root mean square differences of forecasts of surface pressure compared with analyses) of UK Meteorological Office forecasting models for the region around the British Isles since 1966 for 24h, 48h and 72h forecasts.

FIG. 5.6 Illustrating potential improvements in forecast skill if there were better data or a better model (after S. Milton Meteorological Office[4]). The ordinate (vertical axis) is a measure of the error of model forecasts (it is the root mean square differences of forecasts of the 500 hPa height field compared with analyses). Curve (a) is the error of current UK Meteorological Office forecasts as a function of forecast range. Curve (b) is an estimate showing how with the same initial data the error would be reduced if a perfect model could be used. Curve (c) is an estimate showing the further improvement which might be expected if near-perfect data could be provided for the initial state. After a sufficiently long period all the curves approach a saturation value of the average root mean square difference between any forecasts chosen at random.

61

forecast the weather in detail, is there any possibility of predicting the average weather, say, a few months ahead? As this section shows, it is possible for some parts of the world, because of the influence of the distribution of ocean surface temperatures on the atmosphere's behaviour. For seasonal forecasting it is no longer the initial state of the atmosphere about which detailed knowledge is required. Rather, it is the conditions at the surface and how they might be changing which we need to know.

In the tropics, the atmosphere is particularly sensitive to sea surface temperature. This is not surprising because the largest contribution to the heat input to the atmosphere is due to evaporation of water vapour from the ocean surface and its subsequent condensation in the atmosphere, releasing its latent heat. Because the saturation water vapour pressure increases rapidly with temperature—at higher temperatures more water can be evaporated before the atmosphere is

Weather forecasting and chaos[5]

The science of chaos has developed rapidly since the 1960s (when a meteorologist Edward Lorenz was one of its pioneers) along with the power of electronic computers. In this context, chaos is a term with a particular technical meaning. A chaotic system is one whose

behaviour is highly sensitive to the initial conditions from which it started. Even quite simple systems can exhibit chaos under some conditions. For instance, the motion of a simple pendulum (Fig. 5.7) can be 'chaotic' under some circumstances, and because of its extreme sensitivity to small disturbances, its detailed motion is not then predictable.

A condition for chaotic behaviour is that the relationship between the quantities which govern the motion of the system be non-linear, in other words a description of the relationship on a graph would be a curve rather than a straight line[6]. Since the appropriate relationships for the atmosphere are non-linear it can be expected to show chaotic

behaviour. This is illustrated in Fig. 5.6 which shows the improvement in predictability which can be expected if the data describing the initial state are improved. However, even with virtually perfect initial data, the predictability in terms of days ahead only moves from about 6 days to about 20 days, because the atmosphere is a chaotic system.

For the simple pendulum not all situations are chaotic. Not surprisingly, therefore, in a system as complex as the atmosphere, some occasions are more predictable than others. A good illustration of an occasion with particular sensitivity to the initial data and to the way in which the data were assimilated into the model is provided by the

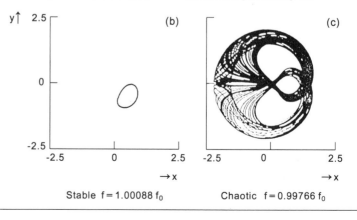

(a)
Forcing frequency f

Conical pendulum (resonant frequency f_0)

y

x

FIG. 5.7 (a) Illustrating a simple pendulum consisting of a bob at the end of a string of length 10cm attached to a point of suspension which is moved with a linear oscillatory forcing motion at frequencies near the pendulum's resonance frequency f_0. The lower diagrams show plots of the bob's motion on a horizontal plane, the scale being in cm. (b) For a forcing frequency just above f_0 the motion of the bob settles down to a simple, regular pattern. (c) For a forcing frequency just below f_0 the bob shows 'chaotic' motion (although contained within a given region) which varies randomly and discontinuously as a function of the initial conditions.[7]

y↑ 2.5

(b)

0

-2.5

-2.5   0   2.5

→ x

Stable f = 1.00088 f_0

(c)

-2.5   0   2.5

→ x

Chaotic f = 0.99766 f_0

saturated—evaporation from the surface and hence the heat input to the atmosphere is particularly large in the tropics.

It is during El Nino events in the east tropical Pacific (see Fig 1.3) that the largest variations are found in ocean temperature. Anomalies in the circulation and rainfall in all tropical regions and to a lesser extent at mid-latitudes are associated with these El Nino events. A good test of the atmospheric models described above is to run them with an El Nino sequence of sea surface temperatures and see whether they are able to simulate these climate anomalies. This has now been done with a number of different atmospheric models; they have shown considerable skill in the simulation of many of the observed anomalies, especially those in the tropics and sub-tropics[8].

Because of the large heat capacity of the oceans, anomalies of ocean surface temperature tend to persist for some months. The

---

Meteorological Office forecasts for the storm which hit Southeast England in the early hours of Friday, 16 October 1987. During this storm gusts of over 90 knots were recorded and approximately 15 million trees were blown down. Although as early as the previous Sunday forecasts had given good early warning of a storm of unusual severity, the model forecasts available during 15 October gave much poorer guidance than earlier forecasts and failed to predict the intensity or the correct track of the storm. The question was raised at the time as to whether the numerical models used were capable of the accurate prediction of such an exceptional event. Fig. 5.8 shows that, using all the data which could have been available and better assimilation procedures in the model, a good forecast of the event can be achieved.

FIG. 5.8 Analyses of the surface pressure in millibars (1 mbar = 100 Pa) and forecasts for the storm which passed over southern England on 15/16 October 1987. The fully developed storm is shown in (b) about 4 hours before it reached the London area; (a) is the situation 24 hours earlier from which the storm developed. (c) and (d) are 24-hour forecasts with the Meteorological Office operational fine-mesh model, (c) being the one available at the time and (d) being one produced after the event using more complete data and better assimilation procedures (from Meteorological Office 1987[9]).

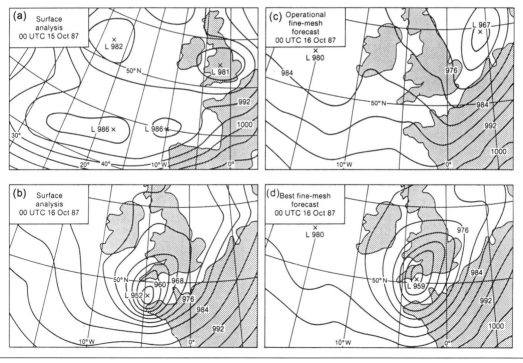

possibility therefore exists, for regions where there is a strong correlation between weather and patterns of ocean surface temperature, of making forecasts of climate (or average weather) some weeks or months in advance. Such seasonal forecasts have been

attempted by scientists from the Meteorological Office for the Sahel region of sub-Saharan Africa, a region where human survival is very dependent on the marginal rainfall (see box).

A further step forward in seasonal forecasting could be taken if

Forecasting for the African Sahel Region

The Sahel region of Africa has been, on average, unusually dry during the past twenty years (Fig. 5.9). This dryness shows a strong correlation with global patterns of ocean surface temperature (Fig. 5.10), a correlation which can be well simulated in runs of global models of the atmospheric circulation similar to weather forecasting models. Fig. 5.11 shows the results of model simulations of the rainfall during the summer months of July to September for a number of years including years both wetter and drier than average. Good agreement is found between simulated and observed rainfall when the model includes ocean surface temperatures for the appropriate periods. Because, however, there are significant variations of ocean surface temperature over a period of months, the agreement becomes poorer when ocean surface temperatures for earlier months are employed in the model, which would have to be the case if the model were used for prediction. The provision of reliable forecasts some months in advance, therefore, will have to await the availability of good forecasts of global ocean surface temperature patterns.

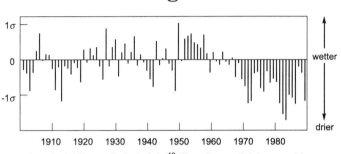

FIG. 5.9 Annual rainfall anomalies in the Sahel[10]. The ordinate scale is in terms of the standard deviation from average.

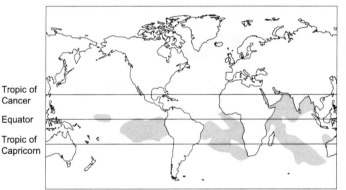

FIG. 5.10 Ocean surface temperatures: difference between Sahel dry years (average of 1972–73, 1982–84) and Sahel wet years (average of 1950, 1952–54, 1958). Dark shading shows regions with differences greater than 0.5°C, light shading shows regions with differences less than –0.5°C[11].

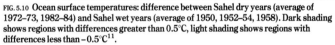

FIG. 5.11 Observed and simulated rainfall in the Sahel[12].

it were possible to forecast the changes in ocean surface temperature. To do that requires understanding of, and the ability to model, the ocean circulation and the way it is coupled to the atmospheric circulation. Because the largest changes in ocean surface temperature occur in the tropics and because there are reasons to suppose that the ocean may be more predictable in the tropics than elsewhere, most emphasis on the prediction of ocean surface temperature has been placed in tropical regions, in particular on the prediction of the El Nino events themselves (see box).

Later on in this chapter the coupling of atmospheric models and ocean models is considered. For the moment it will suffice to say that, using coupled models together with detailed observations of both atmosphere and ocean in the Pacific region, significant skill in the prediction of El Nino events up to a year in advance has been achieved[13] (see also Chapter 7).

The climate system

So far the forecasting of detailed weather over a few days and of average weather for a month or so up to perhaps a season ahead has been described, in order to introduce the science and technology of modelling and also because some of the scientific confidence in the more elaborate climate models arises from their ability to describe and forecast the processes involved in day-to-day weather.

Climate is concerned with substantially longer periods of time, from a few years to perhaps a decade or longer. A description of the climate over a period involves the averages of appropriate components of the weather (for example temperature and rainfall) over that period together with the statistical variations of those components. In considering the effect of human activities such as the burning of fossil fuels, changes in climate over periods of decades up to

---

A simple model of the El Nino

El Nino events are good examples of the strong coupling which occurs between the circulations of ocean and atmosphere. The stress exerted by atmospheric circulation—the wind—on the ocean surface is a main driver for the ocean circulation and, as we have seen, the heat input to the atmosphere from the ocean, especially that arising from evaporation, has a big influence on the atmospheric circulation.

A simple model of an El Nino event which shows the effect of different kinds of wave motions which can propagate within the ocean is illustrated in Fig. 5.12. In this model a wave in the ocean, known as a Rossby wave, propagates westwards from a warm anomaly in ocean surface temperature near the equator. When it reaches the ocean's western boundary it is reflected as a different sort of wave, known as a Kelvin wave, which travels eastward. This Kelvin wave cancels and reverses the sign of the original warm anomaly, so triggering a cold event. The time taken for this half-cycle of the whole El Nino process is determined by the speed with which the waves propagate; it takes about two years. It is essentially driven by ocean dynamics, the associated atmospheric changes being determined by the patterns of ocean surface temperature (and in turn reinforcing those patterns) which result from the ocean dynamics. Expressed in terms of this simple model the El Nino process appears to be essentially predictable.

FIG. 5.12 **Schematic to illustrate El Nino oscillation[14].**

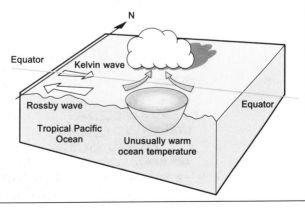

a century or two ahead must be predicted.

Since we live in the atmosphere the variables commonly used to describe climate are mainly concerned with the atmosphere. But climate cannot be described in terms of atmosphere alone. Atmospheric processes are strongly coupled to the oceans (see above); they are also coupled to the land surface. There is also strong coupling to those parts of the Earth covered with ice (the cryosphere) and to the vegetation and other living systems on the land and in the ocean (the biosphere). These five components—atmosphere, ocean, land, ice and biosphere—together make up the climate system (Fig. 5.13).

Feedbacks in the climate system

Chapter 2 considered the rise in global average temperature which would result from the doubling of the concentration of atmospheric carbon dioxide assuming that no other changes occurred apart from the

FIG. 5.13 **Schematic of the climate system**

increased temperature at the surface and in the lower atmosphere. The rise in temperature was found to be 1.2°C. However, it was also established that, because of feedbacks (which may be positive or negative) associated with the temperature increase, the actual rise in global average temperature was likely to be approximately doubled to about 2.5°C. This section lists the most important of these feedbacks.

Water vapour feedback. This is the most important. With a warmer atmosphere more evaporation occurs from the ocean and from wet land surfaces. On average, therefore, a warmer atmosphere will be a wetter one; it will possess a higher water vapour content. Since water vapour is a powerful greenhouse gas, on average a positive feedback results, which on its own would increase the global average temperature rise due to doubled atmospheric carbon dioxide by about 60 per cent.

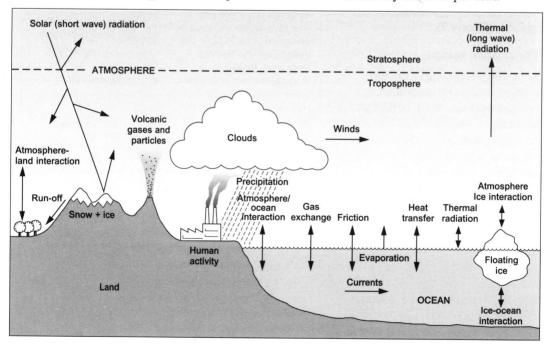

Cloud-radiation feedback. This is more complicated as several processes are involved. Clouds interfere with the transfer of radiation in the atmosphere in two ways (Fig. 5.14). Firstly, they reflect a certain proportion of solar radiation back to space, so reducing the total energy available to the system. Secondly, they act as blankets to thermal radiation from the Earth's surface in a similar way to greenhouse gases. By absorbing thermal radiation emitted by the Earth's surface below, and by themselves emitting thermal radiation, they act to reduce the heat loss to space by the surface.

Which effect dominates for any particular cloud depends on the cloud temperature (and hence on the cloud height) and on its detailed

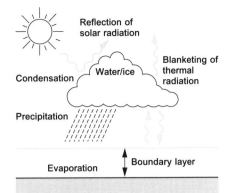

FIG. 5.14 Schematic of the physical processes associated with clouds.

optical properties (those properties which determine its reflectivity to solar radiation and its interaction with thermal radiation). The latter depend on whether the cloud is of water or ice, on its liquid or solid water content (how thick or thin it is) and on the average size of the cloud particles. In general for low clouds

Cloud radiative forcing

A concept helpful in distinguishing between the two effects of clouds mentioned above is that of cloud radiative forcing. Take the radiation leaving the top of the atmosphere above a cloud; suppose it has a value R. Now imagine the cloud to be removed, leaving everything else the same; suppose the radiation leaving the top of the atmosphere is now R'. The difference R'-R is the cloud radiative forcing. It can be separated into solar radiation and thermal radiation. Values of the cloud radiative forcing deduced from satellite observations are illustrated in Fig. 5.15, which shows that on average clouds tend slightly to cool the Earth-atmosphere system.

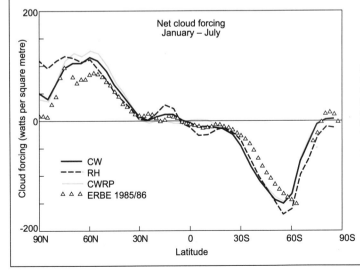

FIG. 5.15 The cloud radiative forcing is made up of a solar radiation and a thermal radiation component which generally act in opposite senses, each typically of magnitude between 50 and 100 watts per square metre. The average net forcing is shown here for the period January to July as a function of latitude as observed from satellites in two years (1985/6) of the Earth Radiative Budget Experiment (ERBE) and also as simulated in a climate model of the Meteorological Office with different schemes of cloud formulation (details and key in box on p 70). There is encouraging agreement between the models' results and observations, but also differences which need to be understood. It is through comparisons of this kind that further elucidation of cloud radiation feedback will be achieved[15].

Global average temperature changes under different assumptions about greenhouse gases and clouds		
Greenhouse gases	Clouds	Change in °C from current average global surface temperature of 15°C
As now	As now	0
None	As now	-32
None	None	-21
As now	None	4
As now	As now but + 3% high cloud	0.3
As now	As now but + 3% low cloud	-1.0
Doubled CO_2 concentration; otherwise as now	As now (no additional cloud feedback)	1.2
Doubled CO_2 concentration + best estimate of feedbacks	Cloud feedback included	2.5

TABLE 5.1 Estimates of global average temperature changes under different assumptions and for changes in greenhouse gases and clouds.

the reflectivity effect wins so they tend to cool the Earth-atmosphere system; for high clouds, by contrast, the blanketing effect is dominant and they tend to warm the system. The overall feedback effect of clouds, therefore, can be either positive or negative.

Climate is very sensitive to possible changes in cloud amount or structure, as can be seen from the results of models discussed in later chapters. To illustrate this, Table 5.1 shows that the hypothetical effect on the climate of changes of a few per cent in cloud cover is comparable with the expected changes due to a doubling of the carbon dioxide concentration.

Ocean-circulation feedback. The oceans play a large part in determining the existing climate of the Earth; they are likely therefore to have an important influence on climate change due to human activities.

The oceans act on the climate in three important ways. Firstly, as we have already noted, they are the main source of atmospheric water vapour which, through its latent heat of condensation in clouds, provides the largest single heat source for the atmosphere.

Secondly, they possess a large heat capacity compared with the atmosphere, in other words a large quantity of heat is needed to raise the temperature of the oceans only slightly. In comparison, the entire heat capacity of the atmosphere is equivalent to less than 3 metres depth of water. That means that in a world which is warming, the oceans warm much more slowly than the atmosphere. We experience this effect of the oceans as they tend to reduce the extremes of atmospheric temperature. For instance, the range of temperature change both during the day and seasonally is much less at places near the coast than at places far inland. The oceans therefore exert a dominant control on the rate at which atmospheric changes occur.

Thirdly, through their internal circulation they redistribute heat throughout the climate system. The total amount of heat transported from the equator to the polar regions by the oceans is similar to that transported by the atmosphere. However, the regional distribution of that transport is very different (Fig. 5.16). Even small changes in the regional heat transport by the oceans could have large implications for climate change. For instance, the amount of heat transported by the north Atlantic Ocean is over 1,000 terawatts (1 terawatt = 1 million million watts = 10^{12} watts). To give an idea of how large this is, we can note that a large power station puts out about 1,000 million (10^9 watts) and the total amount of commercial energy produced globally is about 12 terawatts. To put it further in context, considering the region of the north Atlantic Ocean west of the British Isles, the heat input (Fig. 5.16) carried by the ocean circulation is on average about 100 watts per square metre, of similar magnitude to that reaching the ocean surface there from the incident solar radiation. Any

accurate simulation of likely climate change, therefore, especially of its regional variations, must include a description of ocean structure and dynamics.

Ice-albedo feedback. An ice or snow surface is a powerful reflector of solar radiation (the albedo is a measure of its reflectivity). As some ice melts, therefore, at the warmer surface, solar radiation which had previously been reflected back to space by the ice or snow is absorbed, leading to further increased warming. This is another positive feedback which on its own would increase the global average temperature rise due to doubled carbon dioxide by about 20 per cent.

Four feedbacks have been identified, all of which play a large part in the determination of climate, especially its regional distribution. It is therefore necessary to introduce them into climate models. Because the global models allow for regional variation and also include the important non-linear processes in their formulation, they are able in principle to provide a full description of the effect of these feedbacks. They are, in fact, the only tools available

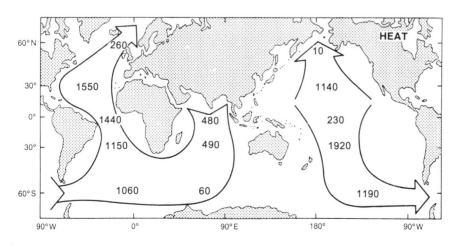

FIG. 5.16 Estimates of transport of heat by the oceans[16]. Units are terawatts (10^{12} W or million million watts). Note the linkages between the oceans and that some of the heat transported by the north Atlantic originates in the Pacific.

69

with this potential capability. It is to a description of climate prediction models that we now turn.

Models for climate prediction

For models to be successful they need to include an adequate description of the feedbacks we have listed. The water vapour feedback and its regional distribution depend on the detailed processes of evaporation, condensation and advection (the transfer of heat by horizontal air flow) of water vapour, and on the way in which convection processes (responsible for showers and thunderstorms) are affected by higher surface temperatures. All these processes are already well included in weather forecasting models, and water vapour feedback has been very thoroughly studied[17]. The most important of the others are the cloud-radiation feedback and the ocean-circulation feedback. How are these incorporated into the models?

Weather-forecasting models and most climate models to date employ comparatively simple schemes for the introduction of cloud. For instance, a typical model will allow for three layers of cloud (low, medium and high), each with prescribed values of reflectivity and transmissivity to radiation. For cloud which results from large-scale atmospheric motions, the presence or absence of cloud in the model is determined by a threshold humidity value; if the humidity is above the threshold value (typically around 90 per cent) cloud is deemed to be present; if the humidity is below that value cloud is absent in the model. The actual threshold value can be adjusted by experiment to ensure that the average cloud cover in the model is approximately that observed in the real atmosphere. For convective cloud, a particular prescription of cloud cover has to be incorporated as part of the model's convection scheme.

A model experiment with different cloud formulations

A model experiment with different cloud formulations has been described by John Mitchell and his colleagues[18] using a climate model at the Hadley Centre at the Meteorological Office. In separate runs of the model he employed three different cloud parametrization schemes:

1 a simple threshold relative humidity scheme as described in the text applied to all clouds (RH);

2 a scheme which includes cloud water as a separate variable and distinguishes between some of the properties of water and ice clouds and the precipitation processes associated with them (CW);

3 as the second scheme, but also allowing the radiative properties of the clouds to be dependent on the cloud water content; in this scheme thick clouds with a large water content are more reflective than thin clouds (CWRP).

In each of the schemes appropriate parameters were adjusted so that the model when run under current conditions reproduced approximately the current distribution of cloud. Fig. 5.15 (see box) compares model results for the three cases with satellite observations.

When the model was run in each case with doubled carbon dioxide concentration, a wide range of results was obtained. For scheme 1, the rise in global average temperature was about 5°C; for scheme 2, about 3°C and for scheme 3, about 2°C. Catherine Senior and John Mitchell point out that although scheme 3 includes the most complete description of the physics of cloud formation, it probably overemphasizes the aspects of negative feedback. They conclude[19] that the climate sensitivity (the rise in global average temperature under equilibrium conditions for an atmosphere with doubled carbon dioxide concentration) is most likely to lie between 2.1°C and 2.8°C, in good agreement with the 'best value' of 2.5°C quoted by the International Panel on Climate Change (IPCC)[20].

The amount and sign (positive or negative) of the average cloud-radiation feedback in the models are dependent on many aspects of a model's formulation as well as on the particular scheme used for the description of cloud formation. Recently more sophisticated schemes for describing cloud formation have been introduced into models (see box). These have given a wide range of results for the average cloud-radiation feedback, some of the feedbacks being positive and some negative. This translates into a range of values for the likely rise in global average temperature for doubled concentration of carbon dioxide which goes from about 5°C for the case with most positive feedback to about 2°C for the one where the feedback is most negative. Comparing the evidence from observations and from models it seems that the most likely average value of cloud-radiation feedback is slightly positive. But there are likely to be important regional variations and our lack of knowledge of this feedback represents the largest uncertainty in model predictions.

The second important feedback is that due to the effects of the ocean circulation. Compared with a global atmospheric model for weather forecasting, the most important elaboration of a climate model is the inclusion of the effects of the ocean. Early climate models only included the ocean very crudely; they represented it by a simple slab some 50 or 100 metres deep, the approximate depth of the 'mixed layer' of ocean which responds to the seasonal heating and cooling at the Earth's surface. In such models, adjustments had to be made to allow for the transport of heat by ocean

currents. But when running the model with a perturbation such as increased carbon dioxide, it was not possible to make allowance for any changes in that transport which might occur. Such models therefore possess severe limitations.

For an adequate description of the influence of the ocean it is necessary to model the ocean circulation and its coupling to the atmospheric circulation. Fig. 5.17 shows the ingredients of such a model. For the atmospheric part of the model, in order to accommodate long runs on available computers, the size of the grid has to be substantially larger, typically 300km in the horizontal. Otherwise it is essentially the same as the global model for weather forecasting described earlier. The formulation of the dynamics and physics of the ocean part of the model is similar to that of its atmospheric counterpart. The effects of water vapour are of course peculiar to the atmosphere, but the salinity (the salt content) of the oceans has to be included as a parameter together with its considerable effects on the water density.

At the ocean-atmosphere interface there is exchange of heat, water and momentum (in this case,

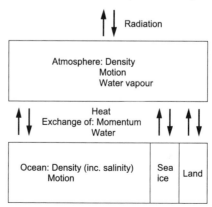

FIG. 5.17 **Component elements and parameters of a coupled atmosphere-ocean model including the exchanges at the atmosphere-ocean interface.**

friction) between the two fluids. The importance of water in the atmosphere in helping to control its circulation has already been shown. The distribution of fresh water precipitated from the atmosphere as rain or snow also has a large influence on the ocean's circulation, because it affects the distribution of salt in the ocean, which in turn affects the ocean density. It is not surprising, therefore, to find that the

'climate' described by the model is quite sensitive to the size and the distribution of water exchanges at the interface.

Before the model can be used for prediction it has to be run for a considerable time until it reaches a steady 'climate'. The 'climate' of the model, when it is run unperturbed by increasing greenhouse gases, should be as close as possible to the current actual climate. If the exchanges are

The ocean's deep circulation

For climate change over periods up to a decade, only the upper layers of the ocean have any substantial interaction with the atmosphere. For longer periods, however, links with the deep ocean circulation become important. The effects of changes in the deep circulation are of particular importance.

Experiments using chemical tracers, for instance those illustrated in Fig. 5.20, have been helpful in indicating the regions where strong coupling to the deep ocean occurs. To sink to the deep ocean, water needs to be particularly dense, in other words both cold and salty. There are two

main regions where such dense water sinks down to the deep ocean, namely in the north Atlantic Ocean between Scandinavia and Greenland and in the region of Antarctica. Salt-laden deep water formed in this way contributes to a deep ocean circulation that involves all the oceans (Fig. 5.18)[21].

An important link exists between this deep ocean circulation and the hydrological (water) cycle. Suppose, for instance, that there is persistent increased precipitation in the north Atlantic region. The surface water will become less salty and therefore less dense. It will not sink so readily and the deep water

formation will be inhibited. This sequence of events has in fact been reported in runs of the climate model at the Geophysics Fluid Dynamics Research Laboratory at Princeton in the USA. As the amount of carbon dioxide in the model was increased, more precipitation occurred over the north Atlantic. With doubled carbon dioxide the rate of deep water formation was reduced by 30 per cent; with quadrupled carbon dioxide, it ceased altogether bringing large changes of climate to that region[22].

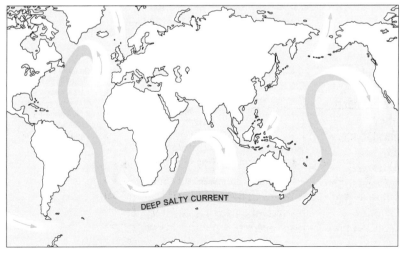

DEEP SALTY CURRENT

FIG. 5.18 Deep water formation and circulation. The deep salty current originates in the north Atlantic. Northward flowing water which is unusually salty becomes cooler and even more salty through evaporation, so increasing its density and causing it to sink[23].

not correctly described, this will not be the case. Getting the exchanges right in the model has proved to be difficult. The current practice is generally to make artificial adjustments to them so as to ensure that the model's 'climate' is as identical as possible to the current climate.

Before leaving the oceans, there is a particular feedback which should be mentioned between the hydrological cycle and the deep ocean circulation (see box). Changes in rainfall, by altering the ocean salinity, can interact with the ocean circulation. This could affect the climate, particularly of the north Atlantic region; it may also have been responsible for some dramatic climate changes in the past[24] (see Chapter 4).

The most important feedbacks belong to the atmospheric and the ocean components of the model. They are the largest components, and because they are both fluids and have to be dynamically coupled together their incorporation into the model is highly demanding. However, another feedback to be modelled is the ice-albedo feedback, which arises from the variations of sea-ice and of snow.

Sea-ice covers a large part of the polar regions in the winter. It is moved about both by the surface wind and by the ocean circulation. So that the ice-albedo feedback can be properly described, the growth, decay and dynamics of sea-ice have to be included in the model. Land ice is also included, essentially as a boundary condition—a fixed quantity—because its coverage changes little from year to year. However, the model needs to show whether there are likely to be changes in ice volume, even though these are small, in order to find out

their effect on sea-level (Chapter 7 considers the impacts of sea-level change).

Interactions with the land surface must also be adequately described. Its most important properties for the model are its wetness or, more precisely, the soil moisture content (which will determine the amount of evaporation) and its albedo (its reflectivity to solar radiation). The models keep track of the changes in soil moisture through evaporation and precipitation. The albedo depends on soil type and vegetation and also on the surface wetness.

Validation of the model

Once a climate model has been formulated it can be tested in three main ways. First, it can be run for a number of years and the climate generated by the model compared in detail to the current climate. For the model to be seen as a valid one, the average distribution and the seasonal variations of appropriate parameters such as surface pressure, temperature and rainfall have to compare well with observation. The movement of chemical tracers can be employed as a test for the ocean component of the model (see box on p 74). In the same way, the variability of the model must be similar to the observed variability. Climate models which are currently employed for climate prediction stand up well to such comparisons[25].

Secondly, models can be compared against simulations of past climates when the distribution of key variables was substantially different than at present, for example the period around 9,000 years ago when the Earth's orbital configuration was different

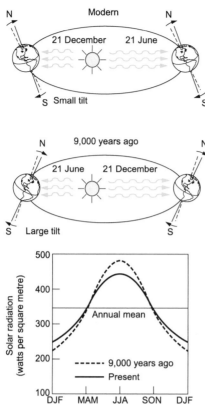

FIG. 5.19 **Changes in the Earth's orbit from the present configuration to 9,000 years ago and (bottom) changes in the average solar radiation during the year over the northern hemisphere**[26].

(see Fig. 5.19). The perihelion (minimum Earth-sun distance) was in July rather than in January as it is now; also the tilt of the Earth's axis was slightly different from its current value (24° rather than 23.5°). Resulting from these orbital differences (see Chapter 4), there were significant differences in the distribution of solar radiation throughout the year. The incoming solar energy when averaged over the northern hemisphere was about 7 per cent greater in July and correspondingly less in January.

When these altered parameters are incorporated into a model, a different climate results. For instance, northern continents are warmer in summer and colder in winter. In summer a significantly expanded low pressure region develops over north Africa and south Asia because of the increased land-ocean temperature contrast. The summer monsoons in these regions

Modelling of tracers in the ocean

A test which assists in validating the ocean component of the model is to compare the distribution of a chemical tracer as observed and as simulated by the model. In the 1950s radioactive tritium (an isotope of hydrogen) released in the major atomic bomb tests entered the oceans and was distributed by the ocean circulation and by mixing. Fig. 5.20 shows good agreement between the observed distribution of tritium (in tritium units) in a section of the western north Atlantic Ocean about a decade after the major bomb tests and the distribution as simulated by a 12-level ocean model.

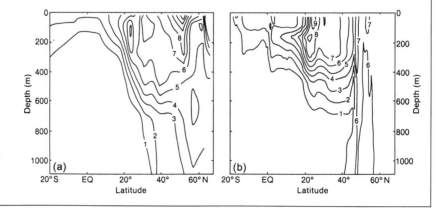

FIG. 5.20 **The tritium distribution in a section of the western north Atlantic Ocean approximately one decade after the major bomb tests; as observed in the GEOSECS programme (a) and as modelled (b)**[27].

are strengthened and there is increased rainfall. These simulated changes are in qualitative agreement with paleoclimate data; for example, these data provide evidence for that period (around 9,000 years ago) of lakes and vegetation in the southern Sahara about 1,000km north of the present limits of vegetation.

The accuracy and the coverage of data available for these past periods are limited. However, the model simulations for 9,000 years ago, described above, and those for other periods in the past have demonstrated the value of such studies in the validation of climate models.

A third way in which models can be validated is to use them to predict the effect of large perturbations on the climate. Good progress is being achieved with the prediction of El Nino events and the associated climate anomalies up to a year ahead (see earlier in the chapter). Other short-term perturbations are due to volcanic eruptions, the effects of which were mentioned in Chapter 1. Several climate models have been run in which the amount of incoming solar radiation has been modified to allow for the effect of the volcanic dust from Mount Pinatubo, which erupted in 1991. Successful simulation of the anomalies of climate which followed that eruption, for instance, the unusually cold winters in the Middle East and the mild winters in western Europe, has been achieved by the models[28].

In these three ways, which cover a range of time-scales, confidence has been built up in the ability of models to predict climate change due to human activities.

Some model results

Having set up a model and validated it against the current climate and against past climates, it can be used for predicting future climates, in particular for predicting the changes in climate which could occur as a result of the increasing concentration of greenhouse gases. The most sophisticated of climate models are those in which the circulations of the atmosphere and the ocean are fully coupled together. Seven centres in the world are currently running such models, two in the USA and one each in the UK, Germany, France, Canada and Australia. Although there are differences between the models, the results from them are broadly similar.

To illustrate the results from these models, we show in Fig. 5.21 the change in hemispheric and global average temperature predicted by one of them during a period when the carbon dioxide concentration in the model's atmosphere was increased by 1 per cent per annum (compound), that is approximately the rate of carbon dioxide increase in the atmosphere predicted under business-as-usual for early next century.

There are a number of interesting features of the results illustrated in Fig. 5.21, as follows[29]:

1 The global average temperature as predicted by the model shows variability of up to a few tenths of a degree Celsius over periods of a few years. This variability is entirely due to internal exchanges in the model between different parts of the climate system. The variations are similar to those which have been observed to occur in the climate during the past 100 years when the climate record has been sufficiently accurate to detect them (see Fig. 3.1).

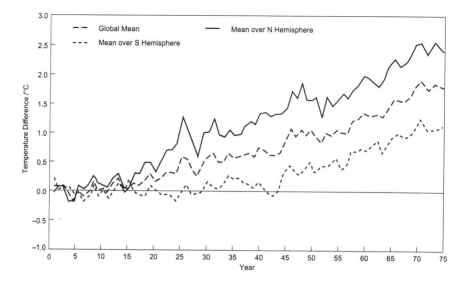

FIG. 5.21 **The change in hemispheric or global average temperature as predicted by the coupled atmosphere-ocean model of the UK Meteorological Office when run with a carbon dioxide concentration increasing at 1 per cent per annum. Over the 70-year period of the model run the carbon dioxide concentration doubles[30].**

2   The relatively slow start to the warming due to the increase in carbon dioxide is probably due to the way the experiment is carried out.

3   The increase of global average temperature after 70 years when the carbon dioxide concentration has doubled is about 1.5°C. This is about 60 per cent of the increase which would occur if the model were run for some time with doubled carbon dioxide concentration so that it could come to equilibrium. The reason for the difference is that the oceans, because of their large thermal capacity, take some time (of the order of twenty years) to warm up.

4   Because of the large influence of the ocean circulation, the warming in the northern hemisphere is almost twice that in the southern hemisphere.

More details of the predictions from this and from other climate models will be presented in the next chapter.

Is the climate chaotic?
Throughout this chapter the implicit assumption has been made that climate change is predictable and that models can be used to provide predictions of climate change due to human activities. Before leaving this chapter I want to consider whether this assumption is justified.

The capability of the models themselves has been demonstrated so far as weather forecasting is concerned. They also possess some skill in seasonal forecasting. They can provide a good description of the current climate and its seasonal variations. Further, they provide predictions which on the whole are reproducible and which are reasonably consistent between different models. But, it might be argued, this consistency could be a property of the models rather than of the climate. Is there any other evidence to support the view that the climate is predictable?

A good place to look for further evidence is in the record of climates of the past, presented in Chapter 4. Correlation between the Milankovitch cycles in the Earth's orbital parameters and the cycles of climate change over the past half million years (see Fig. 4.4, also Fig. 5.19) provides strong evidence to substantiate the Earth's orbital variations as the main factor responsible for the triggering of climate change. The nature of the feedbacks which control the very different amplitudes of response to the three orbital variations still need to be understood. Some 60 ± 10 per cent of the variance in the record of global average temperature from palaeontological sources over the past million years occurs close to frequencies identified in the Milankovitch theory. The existence of this surprising amount of regularity suggests that the climate system is not strongly chaotic so far as these large changes are concerned, but responds in a largely predictable way to the Milankovitch forcing.

This Milankovitch forcing arises from changes in the distribution of solar radiation over the Earth because of variations in the Earth's orbit. Changes in climate as a result of the increase of greenhouse gases are also driven by changes in the radiative regime at the top of the atmosphere. These changes are not dissimilar in kind (although different in distribution) from the changes which provide the Milankovitch forcing. It can be argued therefore that the increases in greenhouse gases will also result in a largely predictable response.

The future of climate modelling

Very little has been said in this chapter about the biosphere. Chapter 3 referred to comparatively simple models of the carbon cycle which include chemical and biological processes and simple non-interactive descriptions of atmospheric processes and ocean transport. The large three-dimensional global circulation climate models described in this chapter contain a lot of dynamics and physics but no interactive chemistry or biology. Detailed scientific knowledge of these areas is not yet sufficient to warrant their inclusion in such complex models; further their inclusion would, at the present, be too demanding in computer time. Eventually, however, it will be important and necessary to have models which are fully comprehensive and interactive and which include chemical and biological processes in the atmosphere, the ocean and on the land.

Climate modelling continues to be a rapidly growing science. Although useful attempts at simple climate models were made with early computers it is only during the last ten years or so that results from them have been sufficiently comprehensive and credible for them to be taken seriously by policy makers, and it is only during the last five years that computers have been powerful enough for coupled atmosphere-ocean models to be employed for climate prediction. A great deal remains to be done to narrow the uncertainty of model predictions. The first priorities must be to improve the modelling of clouds and the description in the models of the ocean-atmosphere interaction. Larger and faster computers are required to tackle this problem, especially to enable the resolution of the model grid to be increased, as well as more sophisticated model physics and dynamics. Much more thorough observations of all components of the climate system are also necessary, so that more accurate validation of the model formulations can be achieved. Very substantial national and international programmes are underway to address these issues.

FOOTNOTES

1 Further information regarding the subject of this chapter can be found in: J.T. Houghton, *The Physics of Atmospheres*, CUP, second edition, 1988; *Climate Change, the IPCC Scientific Assessment*, eds J.T. Houghton, G.J. Jenkins and J.J. Ephraums, CUP, 1990; *Climate Change 1992, the Supplementary Report to the IPCC Scientific Assessment*, eds J.T. Houghton, B.A .Callander and S.K. Varney, CUP, 1992; *Climate System Modelling*, ed. K.E. Trenberth, CUP, 1992; J.T. Houghton, 'The Bakerian Lecture, 1991: 'The predictability of weather and climate', *Phil. Trans. R. Soc. Lond. A (1991)*, **337**, pp 521–72.

2 L.F. Richardson *Weather Prediction by Numerical Processes*, CUP, 1922, reprinted Dover, 1965.

3 For more detail see for instance J.T.Houghton 1988 *loc. cit.*

4 Quoted by J.T.Houghton 1991 *loc. cit.*

5 For more detail see J.T. Houghton 1991 *loc. cit.*; also T.N. Palmer, 'A nonlinear perspective on climate change', *Weather*, **48**, 1993, pp 314–26.

6 An equation such as $y = ax + b$ is linear; a plot of y against x is a straight line. Examples of non-linear equations are $y = ax^2 + b$ or $y + xy = ax + b$; plots of y against x for these equations would not be straight lines. In the case of the pendulum, the equations describing the motion are only approximately linear for very small angles from the vertical where the sine of the angle is approximately equal to the angle; at larger angles this approximation becomes much less accurate and the equations are non-linear.

7 After J. Lighthill, 'The recently recognized failure in Newtonian dynamics', *Proc. Roy. Soc. Lond.* **A407**, 1986, pp 35–50,.

8 For instance, J. Shukla & M.J. Fennessy, *J. Atmos. Sci.*, **45**, 1988 pp 9–28.

9 *The storm 15/16 October 1987*, Meteorological Office Report, Bracknell.

10 Data from S.E. Nicholson and C.K. Folland, see J.T. Houghton 1991 *loc. cit.*

11 C.K. Folland *et al.* 'Sahel rainfall and worldwide sea temperature 1901–1985', *Nature*, **320**, 1986, pp 602–607.

12 M. Hulme *et al.* 'Seasonal rainfall forecasting for Africa: Part 1, Current status and future developments', *Int. J. environ. Stud. A* **39**, 1992, pp 245–56.

13 See for instance M.A. Cane in *Climate System Modelling*, ed. K.E. Trenberth, CUP 1992, Chapter 18, pp 583–616.

14 From J.T. Houghton 1991 *loc. cit.*

15 Diagram from Catherine Senior, Meteorological Office.

16 From J.D. Woods, 'The upper ocean and air sea interaction in global climate' in *The global climate*, ed. J.T. Houghton, CUP, 1984, pp 141–87.

17 R S. Lindzen in a paper 'Some coolness concerning global warming', *Bull. Amer. Met. Soc.*, **71**, 1990, pp 288–99, has suggested that the water vapour feedback could be much less than predicted by models and could even be slightly negative. His suggestions have been reviewed by W.L. Gates *et al.*, 'Climate modelling, climate prediction and model validation' in *Climate Change 1992: the Supplementary Report to the IPCC Assessment*, 1992, pp 97–134, who conclude that the weight of evidence from both observations and models is that the water vapour feedback is positive and of about the magnitude which is predicted by the models.

18 J.F.B. Mitchell, C.A. Senior & W.J. Ingram *Nature*, **341**, 1989, pp 132–34.

19 C.A.Senior and J.F.B.Mitchell 'Carbon dioxide and climate: the impact of cloud parameterization', *J. Climate*, **6**, 1993, pp 393–418.

20 See *Climate Change: the IPCC Scientific Assessment*, 1990, *loc. cit.*

21 A recent review by W.J. Schmitz and M.S. McCartney, 'On the north Atlantic circulation', *Rev. Geophys.* **31**, 1993, pp 29–49, gives details of the north Atlantic circulation and points out the sparse observational evidence for the extent of the deep ocean conveyor as proposed by Broecker and Denton.

22 S. Manabe and R.J. Stouffer, 'Century-scale effects of increased atmospheric CO_2 on the ocean-atmosphere system', *Nature*, **364**, 1993, pp 215–18; details also in *Climate Change: the IPCC Scientific Assessment*, 1990, *op. cit.*, Chapter 6.

23 After W.S. Broecker & G.H. Denton, 'What drives glacial cycles?', *Sci. Amer.* **262**, 1990, pp 43–50.

24 Broecker & Denton, *loc. cit.*

25 For more detail see W.L. Gates, P.R. Rowntree, Q.-C. Zeng, 'Validation of climate models', Chapter 4 in *Climate Change: the IPCC Scientific Assessment*, 1990, pp 93–130.

26 This diagram and information about modelling past climates from J.E. Kutzbach in *Climate System Modelling*, ed. K.E.Trenberth, CUP, 1992, pp 669–701.

27 J.L. Sarmiento, *J.Phys. Oceanog.*, **13**, 1983, 1924–39.

28 H.-F. Graf *et al.* 'Pinatubo eruption winter climate effects: model versus observations,' *Climate Dynamics*, **9**, 1993, pp 61–73.

29 Further details can be found in W.L. Gates *et al.*, 'Climate modelling, climate prediction and model validation' in *Climate Change 1992, the Supplementary Report to the IPCC Scientific Assessment*, eds J.T. Houghton, B.A. Callander and S.K. Varney, CUP, 1992.

30 J.M. Murphy, 'Transient response of the Hadley Centre coupled ocean atmosphere model to increasing carbon dioxide. Part 1: Control climate and flux correction', submitted to *J. Climate*, 1994; and J.M. Murphy and J.F.B. Mitchell, 'Transient response of the Hadley Centre coupled ocean atmosphere model to increasing carbon dioxide. Part 2: Spatial and temporal structure of response', submitted to *J. Climate*, 1994.

6 Climate Change under Business-as-usual

 he last chapter showed that the most effective tool we possess for the prediction of future climate change due to human activities is the climate model. This chapter will describe the predictions of models for likely climate change next century. It will also consider other factors which might lead to climate change and assess their importance relative to the effect of greenhouse gases.

Model predictions

Results of the kind shown in Fig. 5.21 which come from the most sophisticated coupled atmosphere-ocean models available provide fundamental information on which to base climate predictions. However, because they are so demanding on computer time only few results from such models are available. Many studies have also been carried out with simpler models. While these possess a full description of atmospheric processes, they have only a simplified description of the ocean. Much detailed information has come from such models, in spite of their limitations. The predictions presented in the next sections depend on a wide variety of model results including those previously mentioned.

In order to assist comparison between models, experiments with many models have been run with the atmospheric concentration of carbon dioxide doubled from its current level. The global average temperature rise under these conditions of doubled carbon dioxide concentration has become known as the climate sensitivity. The Intergovernmental Panel on Climate Change (IPCC) in its 1990 report gave a 'best estimate' of 2.5°C for the climate sensitivity; it also considered that it was unlikely to lie outside the range of 1.5–4.5°C. The IPCC 1992 Supplementary Report confirmed these values. The predictions presented in this chapter follow the IPCC Assessments[1] and are consistent with these values.

To make predictions of climate change knowledge of future changes in greenhouse gases is first required; this has already been addressed in Chapter 3. The predictions presented in this chapter have used the greenhouse gas emissions scenario IS 92a, published in the IPCC 1992 report (see Fig. 3.5) which, in the absence of any strong controls on emissions is the one based on the

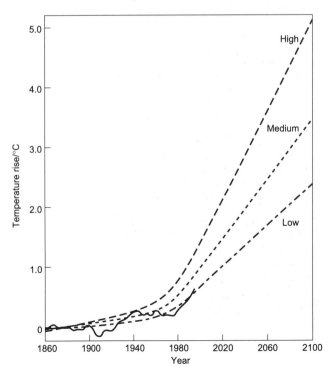

FIG. 6.1 **The predicted change in global average temperature under a 'business-as-usual' scenario (IPCC IS 92a) from 1860 until 2100 (the change is shown from the beginning of pre-industrial times in 1765). The middle curve is the IPCC's best estimate of the change; the upper and lower curves indicate the estimated range of uncertainty. The three curves correspond to 'climate sensitivities' of 4.5, 2.5 and 1.5°C respectively[2]. Also shown is the best estimate of the observed global average temperature from 1860 until the present (see Fig. 3.1).**

most likely assumptions about future conditions. For this reason I have called it the 'business-as-usual' emissions scenario. The method by which the emissions scenario is turned into future projections of greenhouse gas concentrations and into radiative forcing (Fig. 3.8) has been described in Chapter 3.

Predictions of global average temperature

The predicted rise in global average temperature due to the increase in greenhouse gases from pre-industrial times until the year 2100, is shown in Fig. 6.1. From the figure it will be seen that the best estimate of the rise in global average temperature from now until the end of next century is about 2.5°C, or a rate of temperature rise of about a quarter of a degree Celsius per decade. Compared with the temperature changes normally experienced from day to day and

throughout the year, 2.5°C does not seem very much. But, as was pointed out in Chapter 1, it is in fact a large amount when considering a globally averaged temperature. Compare the 5 or 6°C change in global average temperature which occurs between the middle of an ice age and the warm period in between ice ages; 2.5°C is about half an ice-age!

By about the year 2030 when the equivalent amount of carbon dioxide in the atmosphere (see Chapter 3) compared with pre-industrial times will have doubled, the best estimate of the temperature rise to be expected is just under 1°C from now and about 1.6°C from pre-industrial times. As Chapter 5 showed, this is less than the 2.5°C which would be expected for doubled carbon dioxide under steady conditions because of the slowing effect of the oceans on the temperature rise. But this means that by the year 2030, under the business-as-usual scenario, there will be a likely commitment to a temperature rise from pre-industrial times of 2.5°C, even though it will not all have been realized by then.

The rate of change of global average temperature predicted for next century is in the range of 0.15–0.35°C per decade with a best estimate of about 0.25°C per decade. Again this seems a small amount; most people would find it hard to detect a change in temperature of a fraction of a degree. But remembering that these are global averages, these rates of change become very large. Indeed, they are much larger than rates of change for the past few thousand years inferred from paleoclimate data (Fig. 6.2). The ability of ecosystems to adapt to climate change depends critically on the rate of change and for many

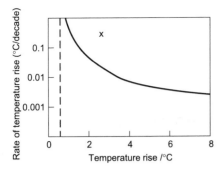

FIG. 6.2 Within the shaded region are typical rates of climate change during the past 10,000 years, over periods up to about a century, plotted against actual change deduced from climate records and paleoclimate data[3]. Also plotted (X) is the predicted rate of change next century due to a business-as-usual increase of greenhouse gases—well outside the shaded region.

over periods of a few years to decades, which probably arises from natural changes within the climate system. The last chapter showed that similar variability occurs in the results of climate models.

Because of the variability, it is difficult to draw any strong conclusions from the trend in the observed record to date. The executive summary of the 1990 IPCC report[4] put the position in a carefully worded statement: 'The size of this (observed) warming is broadly consistent with the predictions of climate models, but it is also of the same magnitude as natural climate variability. Thus the observed increase could be largely due to natural variability; alternatively this variability and other human factors could have offset a still larger human-induced greenhouse warming. The unequivocal detection of the enhanced greenhouse effect from observations is not likely for a decade or more.' In other words, we need to wait a number more years before the global warming signal due to the increase of greenhouse gases stands out clearly above the natural climate variability.

Although I agree with this IPCC statement, it is nevertheless tempting to try to extract indications from the observed record. On inspecting Fig. 6.1 it is clear that a mean curve drawn through this record shows an increase of about 0.5°C compared with a predicted value due to the increase of greenhouse gases of about 0.8°C. This has suggested to some that the models are tending to overestimate greenhouse warming. However, the predictions do not take into account the possible effects of ozone depletion and sulphate dust particles which would both tend to cool the climate.

ecosystems 0.25°C per decade would be a very rapid rate of change indeed.

The curves plotted in Fig. 6.1 go up to the end of the twenty-first century. It is not realistic to suppose, so far into the future, that they have much credibility as predictions. However, they illustrate what is likely to occur if fossil fuels continue to provide most of the world's energy needs. As Chapter 11 will show, there are sufficient reserves of fossil fuels to enable their use to continue to grow well past the year 2100. If that were to happen the global average temperature would continue to rise and could, in the twenty-second century, reach very high levels, perhaps between 5 and 10°C higher than today. The associated changes in climate would be correspondingly large and could well be irreversible.

Comparison with observations

Also shown in Fig. 6.1 is the observed record of global average temperature over the past hundred years or more. As was mentioned in Chapter 3, the most obvious point to note about the record is the significant variability which occurs

Although both ozone change and sulphates show strong variations with latitude and with region, when averaged over the globe their combined effect at the present can be estimated, from the information in Chapter 3, to be a cooling, offsetting perhaps up to one quarter of the warming due to greenhouse gases. Allowing for these effects, therefore, brings the observed and the predicted curves closer together.

If the ozone change and the presence of sulphates have tended so far to reduce the average global temperature increase, will they not tend to do the same in the future? The future projections of Fig. 6.1 and the discussion so far have not allowed for their effects. Should they not have done so? One problem is the uncertainty about how large they are, the level of uncertainty being greater than for the contributions from other greenhouse gases. But that is not the only reason they have not yet been allowed for. Their effects are far from uniform over the globe; ozone depletion is largely in polar regions and sulphate dust particles are largely over the continents in the northern hemisphere. Any changes in climate due to them will be strongly affected by these regional variations, and will therefore be different from that due to the radiative forcing from greenhouse gases, which is more or less uniform over the globe. It is not, therefore, correct to consider their effects as an offset to the climate change due to the increase of the main greenhouse gases, although model calculations including the effects of ozone depletion and sulphate particles need to be made. Further, neither the depletion of stratospheric ozone nor the concentration of sulphates in the atmosphere are expected to become very much larger in the future. Because of the control of CFCs through the Montreal Protocol, ozone depletion should level off about the turn of the century and emissions of sulphur dioxide are also unlikely to increase to a large extent because of concern due to the damage caused by acid rain.

Regional climate change

Discussing changes in global average temperature can produce some kind of overall idea of the magnitude of likely climate change, but in terms of the regional implications, a global average conveys rather little information; regional detail is required. It is in the *regional* or *local* changes that the effects and impacts of global climate change will be felt.

Some very broad regional detail is contained in the results of the coupled atmosphere-ocean model runs described in the last chapter (Fig. 6.3a). The largest increases of temperature occur over the land masses of the northern hemisphere and, because of the important influence of ocean circulation, the smallest temperature increases occur around Greenland and the Norwegian sea and in parts of the Southern oceans.

It is interesting to compare Fig. 6.3a with Fig. 6.3b, which shows the observed change in temperature between the 1980s and the thirty-year period 1950–1980. Although there is a lot of regional variability in the observations—as might be expected, as a decade is a short period over which to take a climate average—there is also a clear similarity in the large-scale patterns of the two figures (for instance the warmer continents and the cool region in the north Atlantic).

This similarity between the recent observed record and model simulations of the temperature change with increased greenhouse gases appears to provide some evidence that the increase in global average temperature in the 1980s results from the increase in greenhouse gases. But I have already expressed caution regarding such a conclusion when the period of comparison is short and the range of natural variability large. Further complications arise when the comparison is carried out in more detail. For instance, inspection of the diurnal variation of temperature over some parts of the world (for instance over the USA. and China) indicates that the increase in average temperature has mainly resulted from increased minimum temperatures at night rather than from increased maximum

FIG. 6.3A The change in the regional distribution of average temperature in °C after 70 years of increasing concentration of atmospheric carbon dioxide at 1% per annum (the concentration has doubled in 70 years) as predicted by the coupled atmosphere-ocean model of the UK Meteorological Office[5].

FIG. 6.3B The observed change in the regional distribution of average temperature between the period 1981–90 and the period 1951–80[6].

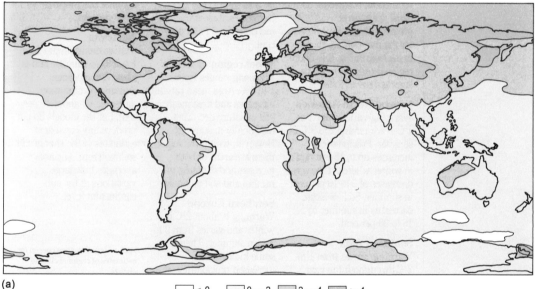

(a)   ☐ < 0   ☐ 0 — 2   ☐ 2 — 4   ☐ > 4

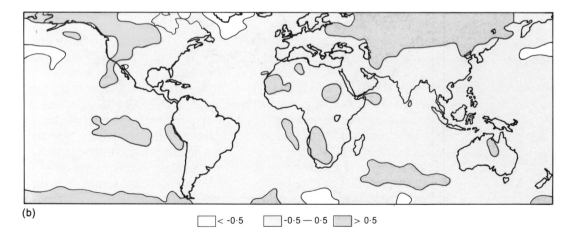

(b)   ☐ < -0·5   ☐ -0·5 — 0·5   ☐ > 0·5

temperatures during daytime[7]— probably due to an increase in cloud cover over these regions. Over other parts of the world (for instance over India) increases of about the same size in both minima and maxima have been recorded. A longer period of comparison between model predictions and observations is required before firmer conclusions can be drawn.

So much for change on the continental scale. Can more specific information be provided about

Estimates of regional changes by 2030[8]

In the IPCC 1990 report, estimates were given for climate changes by the year 2030 under a business-as-usual scenario of greenhouse gas emissions, for the five regions shown in the map of Fig. 6.4. These regional estimates can be summarized as follows:

Central North America
Warming varies from 2 to 4°C in winter and 2 to 3°C in summer. Precipitation increases up to 15 per cent in winter whereas there are decreases of 5 to 10 per cent in summer. Soil moisture decreases in summer by 15 to 20 per cent.

Southern Asia
Warming varies from 1 to 2°C throughout the year.

Precipitation changes little in winter, but in the summer monsoon increases by 5 to 15 per cent[9]. Summer soil moisture increases by 5 to 10 per cent.

Sahel region of Africa
Warming ranges from 1 to 3°C. Area mean rainfall increases and area mean soil moisture decreases marginally in summer. However, within the region, there are areas of both increase and decrease in rainfall and soil moisture.

Southern Europe
Warming is about 2°C in winter and varies from 2 to 3°C in summer. There is some indication of increased precipitation in

winter, but summer precipitation decreases by 5 to 15 per cent, and summer soil moisture by 15 to 25 per cent.

Australia
The warming ranges from 1 to 2°C in summer and is about 2°C in winter. Summer precipitation increases by about 10 per cent, but the models do not produce any consistent estimates of the changes in soil moisture. The area averages hide large variations at the sub-continental level.

FIG. 6.4 **Map of regions chosen for estimates of regional climate change.**

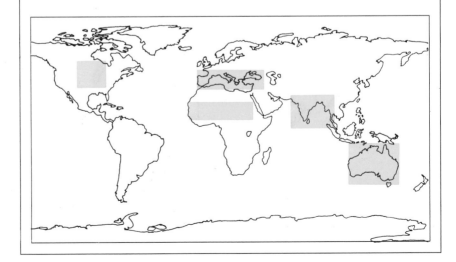

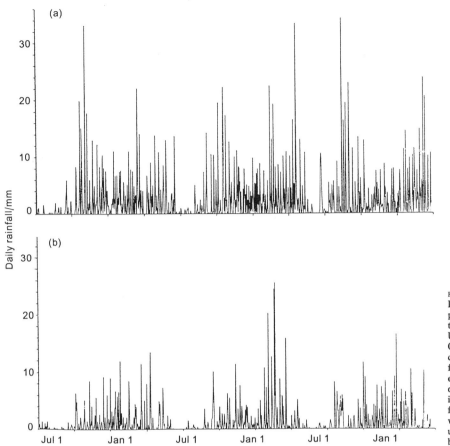

FIG. 6.5 **Daily rainfall for Italy for a three-year period simulated using the climate model at the UK Meteorological Office: (a) for the current climate situation and (b) for the climate if the equivalent carbon dioxide concentration increased by a factor of 4 from its pre-industrial value (predicted to occur under the IS92a scenario by about 2100)[10].**

change for smaller regions? The main difficulty here is the models. When it comes to the detail of regional change, they are inadequate in two main respects. First, the model grid-size in the horizontal is coarse, typically 300km or more, so that climate changes on a smaller scale do not show up. Secondly, as Chapter 5 explains, models deal inadequately with clouds, a problem which particularly influences regional detail. In the IPCC 1990 report, estimates of change for five chosen regions was provided (see box) although, because of the model limitations, it was emphasized that confidence in regional estimates is low. Although many differences in detail in model results remain, modelling research during the past three years has tended to confirm the statements made in 1990. A great deal of research is going on to improve the capability of models to provide regional predictions.

Changes in climate extremes

The last section looked at the likely regional patterns of climate change. Can anything be said about likely changes in the frequency or intensity of climate extremes in the future? It is, after all, not the changes in average climate which are noticeable, but the extremes of climate—the droughts, the floods, the storms and the extremes of temperature in very

85

cold or very warm periods—which provide the largest impact on our lives, as discussed in Chapter 1.

Take, for instance, the likelihood of drought in regions such as Central America or southern Europe, where the average summer rainfall is expected to fall by 10 to 20 per cent. The likely result of such a drop in rainfall is not that the number of rainy days will remain the same, with less rain falling each time. It is more likely that there will be substantially fewer rainy days and considerably more chance of prolonged periods of no rainfall at all (see Fig. 6.5); in other words, much more likelihood of drought. The proportional increase in the likelihood of drought is greater than the proportional decrease in average rainfall.

A similar story emerges when considering what might occur in regions of increased rainfall. Often in such regions the larger amounts of rainfall will come from increased convective activity: more really heavy showers and more intense thunderstorms. The result of a model study of the effect on rainfall in Australia with doubled carbon dioxide is shown in Fig. 6.6. The total rainfall was not much changed, but the number of days with small amounts of rain decreased while the

number of days with large rainfall (greater than 25mm) doubled. The study suggests that the probability of conditions leading to floods would at least have doubled. A similar result (fewer rainy days, higher maximum daily rainfalls for a given mean rainfall rate) has been obtained from the results of the U.K. Meteorological Office climate model[12].

Thus in the warmer world of increased greenhouse gases, different places will experience more frequent droughts and floods. What about other climate extremes, intense storms, for instance? How about hurricanes and typhoons, the violent rotating cyclones which are found over the tropical oceans and which cause such devastation when they hit land? The energy for such storms largely comes from the latent heat of the water which has been evaporated from the warm ocean surface and which condenses in the clouds within the storm, releasing energy. It might be expected that warmer sea temperatures would mean more energy release, leading to more frequent and intense storms. However, ocean temperature is not the only parameter controlling the genesis of tropical storms; the nature of the overall atmospheric flow is also important. Models can take all these factors into account but, because of the relatively large size of their grid, they are unable to simulate very well the detail of relatively small disturbances like tropical cyclones. Model predictions of the changes to be expected in the frequency and intensity of tropical cyclones must therefore be treated with some caution; indeed, simulations of tropical cyclones from different models have produced quite different results[13]. However, a recent study

FIG. 6.6 **Changes in the frequency of the occurrence of different daily rainfall amounts for Australia with doubled carbon dioxide as estimated by the CSIRO climate model in Australia**[11].

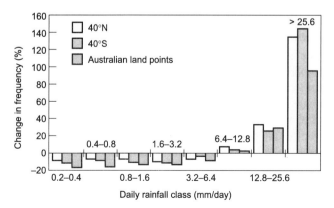

with the Meteorological Office model tends to confirm the simple analysis based on energy considerations that, under conditions with doubled carbon dioxide concentration, there is likely to be an increase (of about 50 per cent according to the study) in the total number of tropical cyclones, together with a relative increase in the proportion of storms of high intensity[14].

The various factors which control the incidence of storms at mid-latitudes are complex. Two factors tend to an increased intensity of storms. The first, as with tropical storms, is that higher temperatures, especially of the ocean surface, tend to lead to more energy being available. The second factor is that the larger temperature contrast between land and sea, especially in the northern hemisphere, tends to generate steeper temperature gradients, which in turn generate stronger flow and greater likelihood of instability. The region around the Atlantic seaboard of Europe is one area where such increased storminess might be expected, a result for which there is some confirmation from model simulations[15]. Much careful study with climate models is going on, aimed at the improvement of understanding and confidence in the detailed predictions of future change, especially those concerned with extremes.

Does the sun's output change?

Some scientists have tried to invoke actual changes in the sun's output to explain climate variations, even short-term ones. Such suggestions are bound to be speculative because no measurements of such changes exist. Accurate measurements of solar output have only been available since 1978, from satellites outside the disturbing effects of the Earth's atmosphere. These measurements indicate a very constant solar output, changing by about 0.1 per cent between a maximum and a minimum in the cycle of solar magnetic activity indicated by the number of sunspots.

It is known from astronomical records and from measurements of radioactive carbon in the atmosphere that this solar sunspot activity has, from time to time over the past few thousand years, shown large variations. Attempts have been made to correlate these sunspot variations with changes found in the historic climate record, such as the Little Ice Age during the seventeenth and eighteenth centuries, and with earlier changes found in the paleoclimate record. Although there are some striking coincidences the overall correlation is not very convincing. Careful studies have estimated that the maximum variations in solar output, if they occur at all, are unlikely to be greater than about 0.5 watts per square metre. This is about the same as the change in the energy regime at the Earth's surface due to only about ten years' increase in greenhouse gases at the current rate[16].

Other factors which might influence climate change

So far climate change due to human activities has been considered. Are there other factors, external to the climate system, which might induce change? Chapter 4 showed that it was variations in the incoming solar energy as a result of changes in the Earth's orbit which triggered the ice ages and the major climate changes of the past. These variations are, of course, still going on; what influence are they having now?

Over the past 10,000 years, because of these orbital changes, the solar radiation incident at 60°N in July has decreased by about 35 watts per square metre, which is quite a large amount. But over 100 years the change is only at most a few tenths of a watt per square metre, which is much less than the changes due to the increases in greenhouse gases (remember that doubling carbon dioxide alters the thermal radiation by about 4 watts per square metre—see Chapter 2).

These orbital changes only alter the distribution of incoming solar energy over the Earth's surface; the total amount of energy reaching the Earth is hardly affected by them. Suggestions have been made that the actual energy output of the sun might change with time. There is no direct evidence for this and, even if such changes occur, they are probably very much smaller than changes in the energy regime at the Earth's surface due to the increase in greenhouse gases (see box).

Another influence on climate comes from volcanic eruptions. Their effects, lasting typically a few years, are relatively short-term compared with the much longer-term effects of the increase of greenhouse gases. The recent large volcanic eruption of Mount Pinatubo in the Philippines which occurred in June 1991 has already been mentioned. Estimates of the change in the net amount of radiation (solar and thermal) at the top of the atmosphere resulting from this eruption are of about 0.5 watts per square metre. This perturbation lasts for about two or three years while the major part of the dust settles out of the atmosphere; the longer term change in forcing, due to the minute particles of dust which last in the stratosphere for much longer, is much smaller.

To summarize this chapter:

■ The increase in greenhouse gases is by far the largest of the factors which can lead to climate change during next century.

■ The likely climate change for a business-as-usual scenario of greenhouse gas emissions has been described in terms of global average temperature and in terms of regional change of temperature and precipitation and the occurrence of extremes.

■ The rate of change is likely to be larger than the Earth has seen at any time during the past 10,000 years.

■ The changes which are likely to have the greatest impact will be changes in the frequencies, intensities and locations of climate extremes—droughts, floods and storms.

The next chapter will look at the impact of such changes on sea-level, on water, on food supplies and on human health. Later chapters of the book will then suggest what action might be taken to slow down and eventually to terminate the rate of change.

FOOTNOTES

1 Climate Change, the IPCC Scientific Assessment, eds J.T. Houghton, G.J. Jenkins and J.J. Ephraums, CUP, 1990; Climate Change 1992, the Supplementary Report to the IPCC Scientific Assessment, eds J.T. Houghton, B.A. Callander and S.K. Varney, CUP, 1992.

2 The information in this diagram is from IPCC 1990 loc. cit. and IPCC 1992 loc. cit. The estimates of temperature rise are based on the radiative forcings shown in Fig. 3.8 which also comes from the IPCC 1992 report. If forcings based on the estimates of carbon dioxide concentration in Fig. 3.6 were used, the estimates of temperature rise would be reduced by about 10 per cent.

3 Adapted from *Climate Change: Meeting the Challenge*, Report by a Commonwealth Group of Experts, Commonwealth Secretariat, London, 1989, p 33 (original source of diagram: Sassin *et al*. 1988).

4 IPCC 1990 *loc. cit.*

5 From J.F.B. Mitchell, Meteorological Office, United Kingdom.

6 From C.K. Folland, Meteorological Office, United Kingdom.

7 C.K. Folland *et al.*, 'Observed Climate Variability and Change', IPCC Supplementary Report 1992, pp 135–170.

8 From IPCC 1990 *loc. cit.*, Policymakers' summary.

9 A recent model result from B. Bhaskaran and J.R. Lavery at the Meteorological Office shows a 30 per cent increase.

10 From C.A. Wilson and J.F.B. Mitchell, 'Simulation of climate and CO_2 induced climate changes over Western Europe', *Climatic Change*, **10**, 1993, pp 11–42.

11 From A.B. Pittock *et al*. 1991, quoted in IPCC 1992 *loc. cit.*, p120.

12 Information from Jonathan Gregory, Hadley Centre, Meteorological Office.

13 J.F.B. Mitchell *et al.*, 'Equilibrium climate change—and its implications for the future' in *Climate Change, the IPCC Scientific Assessment*, eds J.T. Houghton, G.J. Jenkins and J.J. Ephraums, CUP, 1990, pp 131–170.

14 R.J. Haarsma, J.F.B. Mitchell, C.A. Senior, 'Tropical disturbances in a GCM', *Climate Dynamics*, **8**, 1993, pp 247–257.

15 Result from the Meteorological Office model communicated by P.R. Rowntree and C.D. Hall *et al.*

16 I.S.A. Isaksen *et al.*, 'Radiative forcing of climate' in *Climate change 1992, the supplementary report to the IPCC Scientific Assessment*, eds J.T. Houghton, B.A. Callander and S.K. Varney, CUP, 1992, pp 47–67.

7 *The Impacts of Climate Change*

The last two chapters have been detailing the climate change which we can expect next century because of human activities in terms of temperature and rainfall. To be useful to human communities, these details need to be turned into descriptions of the impact of climate change on human resources and activities. The questions to which we want answers are: how much will sea level rise and what effect will that have; how much will water resources be affected; what will be the impact on agriculture and food supply; will human health be affected? This chapter considers these questions.

A complex network of changes

The answers to these questions are far from simple. It is relatively easy to consider the effects of a particular change (in say sea level or water resources) supposing nothing else changes. But other factors will change. Ecosystems have great potential to adapt, and human communities have even more ability to respond and adapt to change. In determining how serious the effects of global warming are likely to be, allowance must be made for response and adaptation. The likely costs of adaptation also need to be taken into account.

The assessment of the impact of global warming is also made more complex because global warming is not the only human induced environmental problem. The loss of soil and its impoverishment (through poor agricultural practice), the over-extraction of groundwater and the damage due to acid rain are examples of environmental degradations on local or regional scales which are having a substantial impact now[1]. If they are not corrected they will tend to exacerbate the negative impacts likely to arise from global warming.

For these reasons, the various effects of climate change so far as they concern human communities and their activities will be put in the context of other factors which might alleviate or exacerbate their impact. The following paragraphs will look at various impacts in turn and then bring them together in a consideration of the overall impact.

How much will sea level rise?

There is plenty of evidence for large changes in sea level during the Earth's past history. For instance, during the warm period before the onset of the last ice age, about 120,000 years ago, the global average temperature was a little warmer than today (Fig. 4.4). Average sea level then was about 5 or 6 metres higher than it is today. When ice cover was at its maximum towards the end of the ice age, some 18,000 years ago, sea level was over 100m lower than today, sufficient, for instance, for Britain to be joined to the continent of Europe.

It is often thought that the main cause of these sea-level changes was the melting or growth of the large ice-sheets that cover the polar regions. It is certainly true that the main reason for the drop in sea level 18,000 years ago was the amount of water locked up in the large extension of the polar ice-sheets. In the northern hemisphere these extended in Europe as far south as southern England and in North America to south of the Great Lakes. It must also be true that the main reason for the 5 or 6m higher sea level during the last warm interglacial period was a reduction in the Antarctic or Greenland ice-sheets. But changes over shorter periods are largely governed by other factors which combine to produce a significant effect on the average sea level.

Various contributions to the likely sea-level rise next century are shown in Fig. 7.1. The largest of these comes from thermal expansion of water in the oceans; as the oceans warm the water expands and the sea-level rises (see box). The other main contribution comes from the melting of glaciers. If all glaciers outside Antarctica and Greenland were to melt, the rise in sea-level would be about 50cm (between 30 and 50cm)[2]. Substantial glacier retreat has occurred during the past century and

Thermal expansion of the oceans

A large component of sea-level rise is due to thermal expansion of the oceans. Calculation of the precise amount of expansion is complex because it depends critically on the water temperature. For cold water the expansion for a given change of temperature is small. The maximum density of sea water occurs at temperatures close to 0°C; for a small temperature rise at a temperature close to 0°C, therefore, the expansion is negligible. At 5°C (a typical temperature at high latitudes), a rise of 1°C causes an increase of water volume of about 1 part in 10,000 and at 25°C (typical of tropical latitudes) the same temperature rise increases the volume by about 3 parts in 10,000. For instance, if the top hundred metres of ocean (which is approximately the depth of what is called the mixed layer) were at 25°C, a rise to 26°C would increase its depth by about 3 centimetres.

A further complication is that not all the ocean changes temperature at the same rate. The mixed layer fairly rapidly comes into equilibrium with changes induced by changes in the atmosphere. The rest of the ocean changes comparatively slowly; some parts may not change at all. To calculate the effect due to thermal expansion, therefore, it is necessary to employ the results of an ocean model—of the kind described in Chapter 5.

FIG. 7.1 **Estimates of sea-level rise next century for the IPCC IS 92a scenario of greenhouse gas emissions (I have called it the business-as-usual scenario), showing the contributions from various factors**[3]**. The uncertainty in these estimates is considered to be about a factor of 2 either way.**

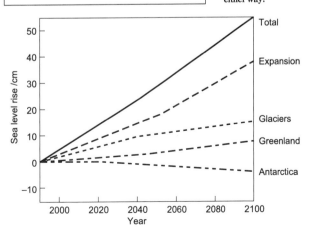

it is estimated that glacier melting may have contributed about 5cm to the observed global sea-level rise of 10–15cm during this period. Modelling the effect of climate change on the behaviour of glaciers is, however, complex. The growth or decay of a glacier depends on the balance between the amount of snowfall on it, especially in winter, and the amount of melting in the summer. Both winter snowfall and average summer temperature are important, and both must be taken into account in future projections of the rate of glacier melting.

It is interesting and perhaps surprising that the net contribution expected from changes in the Antarctic and Greenland ice-sheets is small. For both ice-sheets there are two competing effects[4]. In a warmer world, there is more water vapour in the atmosphere which leads to more snowfall. But there is also more ablation (erosion by melting) of the ice around the boundaries of the ice-sheets where melting of the ice and calving of icebergs occur during the summer months. For Antarctica, the estimates are that accumulation is greater than ablation, leading to a small net growth. For Greenland, ablation is greater than accumulation. For the two taken together the net effect is about zero, although there is considerable uncertainty in that estimate.

A part of the Antarctic ice-sheet in the west of Antarctica is often mentioned as of special concern. A large portion of it is grounded well below sea level. There have been suggestions that it could disintegrate rapidly, in which case its melting would cause sea level to rise by about 5m. It is not known whether melting of the West Antarctic ice-sheet contributed to the rise in sea level during the last interglacial period, 120,000 years ago. Although scientists are not yet very confident in their ability to model the dynamic behaviour of large ice-sheets, there is no reason to suppose there is danger in the short term (for instance, during the next century) of the collapse of any of the major ice-sheets. Much greater understanding of the behaviour of large ice-sheets must be obtained before the amount of warming which might induce such collapse can be estimated.

According to the estimates for the business-as-usual scenario shown in Fig. 7.1, the total average sea-level rise is predicted to be about 15cm by 2030 and about 50cm by the year 2100. Because of the slow rate of rise in temperature of much of the oceans, sea-level rise resulting from global warming will lag behind temperature change at the surface. To illustrate this, Fig. 7.2 shows what would happen if the temperature rose as in the business-as-usual scenario until the year 2030, at which time greenhouse gas concentrations were stabilized so that no further change in radiative forcing of the climate would occur after that date. By 2030, sea level would have risen about 15cm. During the remaining 70 years of the next century a further 20cm or so of sea-level rise would be expected, the rise continuing through the following century too.

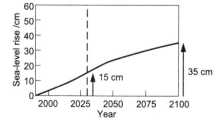

FIG. 7.2 **Estimate of sea-level rise under the IPCC business-as-usual scenario of greenhouse gas emissions until the year 2030. An additional rise in sea level will occur during the remainder of the century even if climate forcing is stabilized in 2030[5].**

These estimates of average sea-level rise provide a general guide as to what can be expected. Sea-level rise, however, will not be uniform over the globe. The effects of thermal expansion in the oceans will vary considerably with location. Further, movements of the land occurring for natural reasons (for instance, the relief of stress in the Earth's mantle due to the removal of the last ice-sheets, which still continues), or because of human activities (for instance, the removal of groundwater), can have comparable effects to the rate of sea-level rise arising from global warming. At any given place, all these factors have to be taken into account in determining the likely value of future sea-level rise.

The impacts of sea-level rise

A rise in average sea level of 15cm by 2030 and about half a metre by the end of the next century may not seem a great deal. Many people live sufficiently above the level of high water not to be directly affected. However, half of humanity inhabits the coastal zones around the world. Within these, the lowest lying are some of the most fertile and densely populated. To people living in these areas, even a fraction of a metre increase in sea level can add enormously to their problems. Some of the areas which are especially vulnerable are Bangladesh and similar delta areas, the Netherlands and the small low-lying islands in the Pacific and other oceans.

Bangladesh is a densely populated country of about 120 million people located in the complex delta region of the Ganges, Brahmaputra and Meghna Rivers. About 7 per cent of the country's habitable land (with about 6 million

population) is less than 1 metre above sea level and about 25 per cent (with about 30 million population) is below the 3 metre contour (Fig. 7.3). Estimates of the sea-level rise next century are of about 1 metre by 2050 (compounded of 70cm due to subsidence because of land movements and removal of groundwater and 30cm from the effects of global warming) and nearly 2m by 2100 (1.2m due to subsidence and 70cm from global warming)[6]— although there is a large uncertainty in these estimates.

It is quite impractical to consider full protection of the long and complicated coastline of Bangladesh from sea-level rise. Its most obvious effect, therefore, is that substantial amounts of good agricultural land

FIG. 7.3 **Land affected in Bangladesh by various amounts of sea-level rise**[7].

will be lost. This is serious: half the country's economy comes from agriculture. 85 per cent of the nation's population depends on agriculture for its livelihood, and many of these people are at the very edge of subsistence.

But the loss of land is not the only effect of sea-level rise. Bangladesh is extremely prone to damage from storm surges. Every year, on average, at least one major cyclone attacks Bangladesh. During the past twenty-five years there have been two very large disasters with extensive flooding and loss of life. The storm surge in November 1970 is probably the largest of the world's natural disasters in recent times; it is estimated to have claimed the lives of over a quarter of a million people. Well over a hundred thousand are thought to have lost their lives in a similar storm in April 1991. Even small rises in sea level add to the vulnerability of the region to such storms.

There is a further effect of sea-level rise on the productivity of agricultural land: the intrusion of saltwater into fresh groundwater resources. At the present time, it is estimated that in some parts of Bangladesh saltwater extends seasonally inland over 150km. With a one metre rise in sea level, the area affected by saline intrusion could double, with a large effect on the supply of fresh water[8].

What possible responses can Bangladesh make to these likely future problems? Over the time-scale of change which is currently envisaged it can be supposed that the fishing industry can relocate and respond with flexibility to changing fishing areas and changing conditions. It is less easy to see what the population of the affected

agricultural areas can do to relocate or to adapt. No significant areas of agricultural land are available elsewhere in Bangladesh to replace that lost to the sea, nor is there anywhere else in Bangladesh where the population of the delta region can easily be located. It is clear that very careful study and management of all aspects of the problem is required. The sediment brought down by the rivers into the delta region is of particular importance. The amount of sediment and how it is used can have a large effect on the level of the land affected by sea-level rise. Careful management is therefore required upstream as well as in the delta itself; groundwater as well as sea defences must also be managed carefully if some alleviation of the effects of sea-level rise is to be achieved.

A similar situation exists in the Nile delta region of Egypt. The likely rise in sea level next century is made up from local subsidence and from global warming in much the same way as for Bangladesh—approximately 1 metre by 2050 and 2 metres by 2100. About 12 per cent of the country's arable land with a population of over 7 million people would be affected by a 1 metre rise of sea level[9]. Some protection from the sea is afforded by the extensive sand dunes but only up to half a metre or so of sea-level rise[10].

Many other examples of vulnerable delta regions, especially in south-east Asia and Africa, can be given where the problems would be similar to those in Bangladesh and in Egypt. For instance, several large and low-lying alluvial plains are distributed along the eastern coastline of China. A sea-level rise of just half a metre would inundate an area of about 40,000 square kilometres (about the area of the Netherlands)[11] where over 30 million people currently live. A particular delta which has been extensively studied is that of the Mississippi in North America. These studies underline the point that human activities and industry are already exacerbating the potential problems of sea-level rise due to global warming. Because of river management little sediment is delivered by the river to the delta to counter the subsidence occurring because of long-term movements of the Earth's crust. Also, the building of canals and dykes has inhibited the input of sediments from the ocean[12]. Studies of this kind emphasize the importance of careful management of all activities influencing such regions, and the necessity of making maximum use of natural processes in ensuring their continued viability.

We now turn to the Netherlands, a country more than half consisting of coastal lowlands, mainly below present sea level. It is one of the most densely populated areas in the world; 8 of the 14 million inhabitants of the region live in large cities like Rotterdam, The Hague and Amsterdam. An elaborate system of about 400km of dykes and coastal dunes, built up over many years, protects it from the sea. Recent methods of protection, rather than creating solid bulwarks, make use of the effects of various forces (tides, currents, waves, wind and gravity) on the sands and sediments so as to create a stable barrier against the sea—similar policies are advocated for the protection of the Norfolk coast in eastern England[13]. Protection against sea-level rise next century will require no new technology.

Dykes and sand dunes will need to be raised; additional pumping will also be necessary to combat the incursion of salt water into freshwater aquifers. It is estimated[14] that an expenditure of about 10 thousand million dollars would be required for protection against a sea-level rise of 1 metre.

The third type of area of especial vulnerability is the low-lying small island. Half a million people live in archipelagos of small islands and coral atolls, such as the Maldives in the Indian Ocean, consisting of 1190 individual islands, and the Marshall Islands in the Pacific, which lie almost entirely within three metres of sea level. Half a metre or more of sea-level rise would reduce their areas and remove up to 50 per cent of their groundwater. The cost of protection from the sea is far beyond the resources of these islands' populations. For coral atolls, however, there is reasonable hope that coral growth, providing it is undisturbed, will match sea-level rise up to the rate of about half a metre a century[15].

These are some examples of areas particularly vulnerable to sea-level rise. Many other areas in the world will be affected in similar, although perhaps less dramatic, ways. Many of the world's cities are close to sea level and are being increasingly affected by subsidence because of the withdrawal of groundwater. The rise of sea level due to global warming will add to this problem. There is no technical difficulty for most cities in taking care of these problems, but the cost of doing so must be included when calculating the overall impact of global warming.

So far, in considering the impact of sea-level rise, places of dense population where there is a large effect on people have been considered. There are also areas of importance where few people live. The world's wetlands and mangrove swamps currently occupy an area of about a million square kilometres (the figure is not known very precisely), equal approximately to twice the area of France. Their biological productivity equals or exceeds that of any other natural or agricultural system. Over two-thirds of the fish caught for human consumption, as well as many birds and animals, depend on coastal marshes and swamps for part of their life cycles, so they are vital to the total world ecology. Such areas can adjust to slow levels of sea-level rise, but there is no evidence that they could keep pace with a rate of rise of greater than about 2mm per year—20cm per century. What will tend to occur, therefore, is that the area of wetlands will extend inland, often with a loss of good agricultural land. Sea defences could be erected to prevent this, but at considerable cost and with a net loss of wetland area[16].

To summarize the impact of the half metre or so of sea-level rise due to global warming which could occur next century: global warming is not the only reason for sea-level rise but it is likely to exacerbate the impacts of other environmental problems. Careful management of human activities in the affected areas can do a lot to alleviate the likely effects, but substantial adverse impacts will remain. In delta regions, which are particularly vulnerable, sea-level rise will lead to substantial loss of agricultural land and salt intrusion into fresh water resources. In Bangladesh, for instance, over 10 million people are likely to be affected by such loss. A further problem in Bangladesh and other low-lying tropical areas will be the increased intensity of disasters because of storm surges. Low-lying small islands will also suffer loss of land and fresh water supplies. Countries like the Netherlands and many cities in coastal regions will have to spend substantial sums on protection against the sea. Significant amounts of land will also be lost near the important wetland areas of the world. Attempts to put costs against these impacts, in both money and human terms, will be considered later in the chapter.

The impact on fresh water resources

The global water cycle is a fundamental component of the climate system. Water is cycled between the oceans, the atmosphere and the land surface (Fig. 7.4). Through evaporation and condensation it provides the main means whereby energy is transferred to the atmosphere and within it. Water is essential to all forms of life; the main reason for the wide range of life forms, both plant and animal, on the Earth is the extremely wide range of variation in the availability of water. In wet tropical forests, the jungle teems with life of enormous variety. In drier regions sparse vegetation exists, of a kind which can survive for long periods with the minimum of water; animals there are also well adapted to dry conditions.

Water is also a key substance for humankind; we need to drink it, we need it for the production of food, for health and hygiene, for industry and transport. Humans have learnt that the ways of providing for livelihood can be adapted to a wide variety of circumstances regarding water supply except, perhaps, for the completely dry desert. Water availability averaged per capita in different countries varies from 1,000 cubic metres (220,000 imperial gallons) per year to over 50,000 cubic metres (11 million imperial gallons)[17]—although quoting average numbers of that kind hides the enormous disparity between those in very poor areas who may walk many hours each day to fetch a few gallons and many in the developed world who have access to virtually unlimited supplies at the turn of a tap.

The demands of increased populations and the desire for higher standards of living have brought with them much greater requirements for fresh water. During the last 50 years water use worldwide has grown fourfold (Fig. 7.5); it now amounts to about 10 per cent of the estimated global total of the river and groundwater flow from land to sea (Fig. 7.4). Two-thirds of human water use is currently for agriculture, much of it for irrigation; about a quarter is used by industry; only 10 per cent or so is used domestically. Increasingly, water stored over hundreds or thousands of years in underground aquifers is being tapped for current use. With this rapid growth of demand comes greatly increased vulnerability regarding water supplies.

A further vulnerability arises because many of the world's major sources of water are shared. About

FIG. 7.4 The global water cycle (in thousands of cubic kilometres per year)[18], showing the key processes of evaporation, precipitation, transport as vapour by atmospheric movements and transport from the land to the oceans by run-off or groundwater flow.

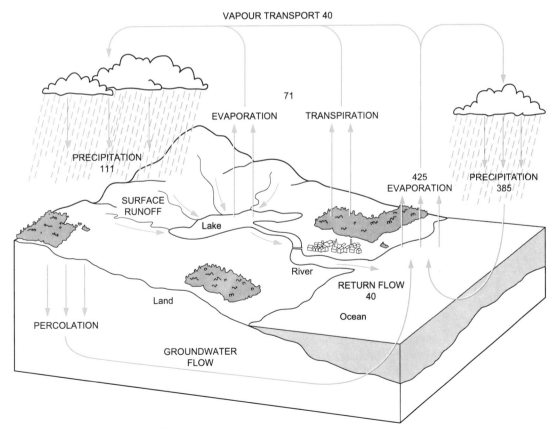

97

half the land area of the world is within water basins which fall between two or more countries. There are 44 countries for which at least 80 per cent of their land areas falls within such international

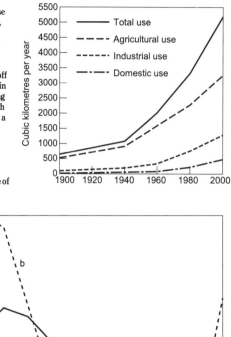

FIG. 7.5 Global water use for different purposes, 1900–2000, in cubic kilometres per year[19].

FIG. 7.6 Simulations of average monthly run-off in the Sacramento basin of California comparing current climate (a) with changed climates with a 4°C temperature increase and 20% increase in rainfall (b) and with the same temperature increase but with 20% decrease of rainfall (c)[20].

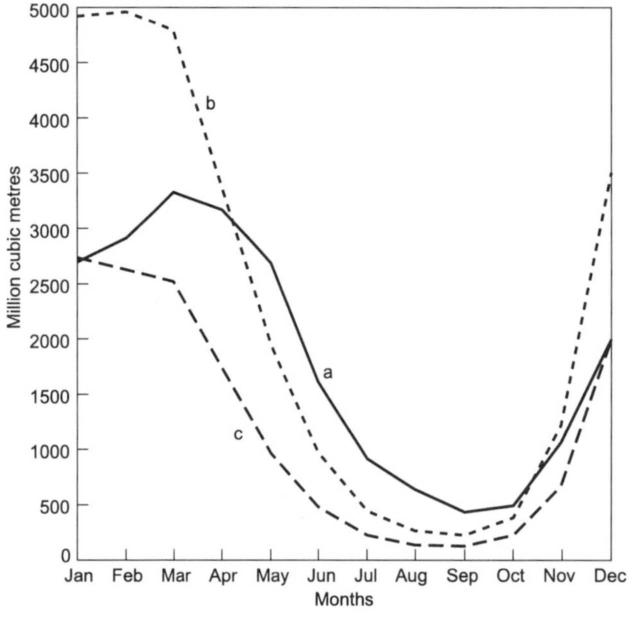

basins. The Danube, for instance, passes through 12 countries which use its water, the Nile through 9, the Ganges-Bramaputra through 5. Other countries where water is scarce are critically dependent on sharing the resources of rivers such as the Euphrates and the Jordan. The

achievements of agreements to share water often bring with them demands for more effective use of the water and better management. Failure to agree brings increased possibility of tension and conflict. The United Nations Secretary-General, Boutros Boutros-Ghali has said that 'The next war in the Middle East will be fought over water, not politics'[21].

The availability of fresh water will be substantially changed in a world affected by global warming. The increase of temperature will mean that a higher proportion of the water falling on the Earth's surface will evaporate. This would not matter if there were more rainfall to make up for the evaporation. However, the predictions presented in Chapter 6, for a business-as-usual scenario of carbon dioxide emissions next century, showed that some parts of the world would have less rainfall, especially in summer. The combined effect of less rainfall and more evaporation in these places would mean less run-off. Other parts of the world would have increased precipitation, for instance the monsoon regions of south-east Asia, where studies indicate substantially greater run-off in the summer (see box on p 102).

The run-off in rivers and streams is what is left from the precipitation which falls on the land after some has been taken by evaporation and by transpiration from plants; it is the major part of what is available for human use. The amount of run-off is highly sensitive to changes in climate; even small changes in the amount of precipitation or in the temperature (affecting the amount of evaporation) can have a big influence on it. To illustrate this Fig. 7.6 shows simulations by Peter Gleick, carried out for the Sacramento Basin in California, USA, of changes in run-off with changed climate conditions. With a 4°C regional temperature rise and 20 per cent decreased rainfall, the run-off in the summer months falls to between 20 per cent and 50 per cent of its normal value. Even with 20 per cent increased rainfall and the same temperature increase, summer run-off still remains well below normal. Watersheds in arid or semi-arid regions are especially sensitive because the annual run-off is in any case highly variable. Some watersheds in mid latitudes in the northern hemisphere, where snowmelt is an important source of run-off, can also be severely affected. For these places, as temperatures rise, winter run-off will increase substantially and spring high water will be much reduced.

So far changes in *average* temperature and *average* rainfall have been discussed; the simulations in Fig. 7.6 are for *average* conditions. But, as has been constantly emphasized, the severity of climate impacts depends to a great degree on extreme conditions. This is well illustrated by looking at the scale of natural disasters involving water— either too much water in floods or too little in droughts. In the list of natural disasters listed in Table 1.1, floods were third (after tropical cyclones and Earthquakes) in order of severity leading to loss of life. A Unesco study in 1973 estimated that each year, in Asia alone, river floods damage or destroy about 40,000 square kilometres of land and crops, and affect the lives and wellbeing of over 17 million people. During the period 1980-85, more than 160 major floods were recorded, killing or injuring more than 120,000 people, destroying the homes of nearly 20 million and causing damage estimated at over 20 thousand million US dollars—a very large amount indeed for the economies of the countries involved. Droughts do not appear high up on the table of natural disasters, not because they are unimportant, but because, unlike most other disasters, their effects tend to felt over a long period of time. The 'dust bowl' years in the 1930s in the United States are still within living memory, as are the droughts and famines in India in 1965–67 which, it is estimated, claimed one and a half million lives. We have all been made very aware of the disastrous consequences of drought in the Sahel region and in other parts of Africa in the 1980s— which are still recurring only too frequently on that continent.

Any temperature or rainfall record shows a large variability. The inevitable result of variability added to higher average temperatures (meaning higher evaporation) and higher average rainfall will be a greater number and greater intensity of both droughts and floods[22]. Some of the areas likely to be affected are just those areas which are particularly vulnerable to floods and droughts at the moment. But droughts and floods are also likely to occur in locations where, at present, such disasters are rare.

We can identify watersheds that are particularly vulnerable to climate change by asking certain questions about them[23].

How much water storage is there in the watershed relative to the annual flow? In Colorado, in the United States, for instance, the storage is four times the annual flow, whereas in the Atlantic States it is only one tenth of the annual flow.

How large is the demand as a percentage of the potential supply? Again in the United States, for the Missouri river basin it is 30 per cent, for the Rio Grande it is 64 per cent and for the lower Colorado 96 per cent. Almost none of the water in the Colorado river reaches the sea. Therefore, though the Colorado has substantial storage and is therefore not very sensitive to annual variations, the amount of use in its lower reaches means that, over a number of years, any reduction of its flow is bound to imply lower water availability.

How much groundwater is being used? There are many places in the world where groundwater is being used faster than it is being replenished. To give two examples, for more than half the land area of the United States over a quarter of the groundwater withdrawn is not being replenished, so every year the water has to be extracted from deeper levels; and in Beijing in China the water table is falling by two metres a year as its groundwater is pumped out.

How variable are the stream and river flows? This question is particularly relevant to arid and semi-arid areas. Detailed studies taking these criteria into account have been carried out for a number of areas; one example for the MINK (Missouri, Iowa, Nebraska and Kansas) region of the United States is shown in the box.

There is another reason, not unconnected with global warming, for the vulnerability of water supplies: the link between rainfall and changes in land use. Extensive deforestation can lead to large changes in rainfall (see box). A similar tendency to reduced rainfall can be expected if there is a reduction in vegetation over large areas of semi-arid regions. Such changes can have a devastating and widespread effect and assist in the process of desertification. This is a potential threat to the drylands covering about

Study of the 'MINK' region in the United States

The United States Department of Energy has carried out a detailed study[24] of the likely effects of climate change on a region (known as the MINK region) in the centre of the United States comprising the states of Missouri, Iowa, Nebraska and Kansas. Included within the region are parts of four major river basins—the Missouri, the Arkansas, the Upper and the Lower Mississippi. Water is already a scarce resource within the MINK states; much of the area's irrigation relies on non-renewable groundwater supplies. These will diminish with time, so that even in the absence of climate change less water will be available, especially for irrigation.

To provide an analogue of the climate which might be expected with increased carbon dioxide, the period of the 1930s was chosen, when the average temperature in the region was about 1°C warmer than in the period 1950–1980 (the 'control' period) and the average precipitation about 10 per cent lower than in the control period.

Water would become scarcer under the analogue climate compared with the control[25]. The hotter and drier conditions would increase evaporation and reduce run-off. Streamflow would drop by about 30 per cent in the Missouri and the Upper Mississippi basins and by about 10 per cent in the

Arkansas. Most streams would fall well short of supplying both the desired instream flows and the current levels of consumption use.

Under the analogue climate, irrigated agriculture would be bound to decline substantially because of the increased constraints on groundwater use coupled with less water availability from other sources. This would also result in a drive to increased efficiency, albeit at greater cost. Maintaining the high priority currently given to navigation on the main stem of the Missouri would become very costly.

Desertification

Drylands (defined as those areas where precipitation is low and where rainfall typically consists of small, erratic, short, high-intensity storms) cover about 40 per cent of the total land area of the world and support over one fifth of the world's population. Desertification in these drylands is the degradation of land because of decreased vegetation, reduction of available water, reduction of crop yields and erosion of soil. It results from excessive land use generally because of increased population, increased human needs, or political or economic pressures (for instance, the need to grow cash crops to raise foreign currency). It is often triggered or intensified by a naturally occurring drought.

The rate of desertification is currently about 60,000 square kilometres per year or 0.1 per cent per year of the total area of drylands[26]. It is a potential threat to 70 per cent of these drylands— over 25 per cent of the world's land area. Fig. 7.7 shows how these arid areas are distributed over the continents.

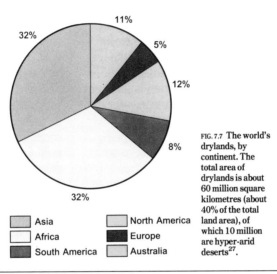

FIG. 7.7 **The world's drylands, by continent. The total area of drylands is about 60 million square kilometres (about 40% of the total land area), of which 10 million are hyper-arid deserts[27].**

Legend:
- Asia
- Africa
- South America
- North America
- Europe
- Australia

Deforestation and changes in rainfall

Changes in land use such as those brought about by deforestation can have an effect on the amount of rainfall, for three main reasons. Over a forest there is a lot more evaporation of water (through the leaves of the trees) than there is over grassland or bare soil, hence the air will contain more water vapour. Also, a forest reflects 12–15 per cent of the sunlight which falls on it, whereas grassland will reflect about 20 per cent and desert sand up to 40 per cent. A third reason arises from the roughness of the surface when vegetation is present.

Jules Charney suggested in 1975, in the context of the drought in the Sahel, that there could be an important link between changes of vegetation (and hence changes of reflectivity) and rainfall. The increased energy absorbed at the surface when vegetation is present and the increased surface roughness both tend to stimulate convection and other dynamic activity in the atmosphere which leads to the production of rainfall.

Experiments with numerical models which include these physical processes demonstrate the effect. They indicate a reduction of about 15 per cent in rainfall if the forest north of 30°S in South America were removed and replaced by grassland[28]. Similar model experiments for Zaire over a smaller region show an average reduction in rainfall of over 30 per cent[29]. A much more drastic experiment in which the Amazonian forest was removed and replaced with a desert surface showed a reduction in rainfall by 70 per cent to levels similar to that of the semi-arid regions of the Sahel part of Africa[30]. Such a model experiment does not represent a realistic situation, but it illustrates the significant impact that widespread deforestation could have on the local climate.

one quarter of the land area of the world (see box).

What sort of action can be taken to decrease the vulnerability of human communities to water supplies? Irrigation accounts for about two-thirds of world water use, and is of great importance to world agriculture. Irrigation is applied to about one-sixth of the world's farmland which produces about one-third of the world's crops. In some areas the ratio is much higher; for instance, over 80 per cent of the agricultural land in California is irrigated. Most irrigation is through open ditches, which is very wasteful of water; over 60 per cent is lost through evaporation and seepage. Microirrigation techniques, in which perforated pipes deliver water directly to the plants, provide large opportunities for water conservation, making it possible to expand irrigated fields without building new dams[31]. Management of the existing infrastructure can be improved, for

instance by arranging for the integration of different supplies, and conservation in the domestic and industrial sectors can be encouraged. Most of these actions will cost money, although they may be much more cost-effective ways of coping with future change in water resources than attempting to develop major new facilities[32].

In summary, what are the likely effects of global warming on water supplies? First, the current vulnerability of many communities to water shortage should be noted. This is especially true of arid and semi-arid regions where the increasing demands of human communities mean that droughts, even for short periods, are more disastrous than before. Vulnerability is well demonstrated in many areas of the world where the amounts of groundwater extraction greatly exceed its replenishment—a situation which cannot continue for very long into the future. Because of

The monsoon regions of south-east Asia

A study for south-east Asia of the change in water resources which might occur with global warming has been carried out by M. Lal from the Centre for Atmospheric Sciences in New Delhi[33]. He has used projections of future climate from the coupled atmosphere-ocean model of the Max Planck Institute in Hamburg, which has been shown to produce good simulations of the precipitation patterns for the region under the current climate.

For the business-as-usual increase in greenhouse gas emissions, the model results indicate for the next hundred years (up to about 2080) an average surface temperature increase over

the land areas of about 3°C, with increased precipitation and run-off in the flood-prone areas of north-east India and south China during the summer (Fig. 7.8). Lal points out

the implications that such changes would have for water management and for land management, especially the control of erosion.[34]

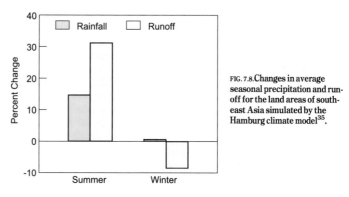

FIG. 7.8. Changes in average seasonal precipitation and run-off for the land areas of south-east Asia simulated by the Hamburg climate model[35].

population growth these vulnerabilities will increase and will exacerbate the negative effects of global warming.

Secondly, climate change because of global warming will result in large changes in water supplies in many places. Although the present state of knowledge regarding regional and local climate change does not allow scientists to identify clearly the most vulnerable areas, they are able to indicate the sort of area which will be most affected. Such areas are those arid and semi-arid areas with reduced rainfall leading to greater aridity and even desertification; continental areas where decreased summer rainfall and increased temperature result in a substantial loss in soil moisture and much increased vulnerability to drought; and areas such as those served by the Asian monsoon, where increased rainfall will lead to a greater incidence of floods. The changing pattern of climate extremes, especially droughts and floods, will be the cause of most of the problems. It is also the case that regions such as south-east Asia that are dependent on unregulated river systems are more sensitive to change than regions such as western Russia and the western United States that have large, regulated water resource systems[36]. Thirdly, some alleviating actions that can be taken have been described; the possible cost of these will be considered later in the chapter.

Impact on agriculture and food supply

Every farmer understands the need to grow crops or rear animals which are suited to the local climate. The distribution of temperature and rainfall during the year are key factors in making decisions regarding what crops to grow. These will change in the world influenced by global warming. The patterns of what crops are grown where will therefore also change. But these changes will be complex; economic and other factors will take their place alongside climate change in the decision-making process.

There is enormous capacity for adaptation in the growth of crops for food. We have seen some of this with what was called the Green Revolution of the 1960s when the development of new strains of many species of crops resulted in large increases in productivity. Between the mid-1960s and the mid-1980s global food production rose by an average annual rate of 2.4 per cent,—faster than global population—more than doubling over that thirty-year period. Grain production grew even faster, at an annual rate of 2.9 per cent[37].

With the detailed knowledge of the conditions required by different species and the expertise in genetic

The carbon dioxide fertilization effect

An important positive effect of increased carbon dioxide concentrations in the atmosphere is the boost to fertilization in plants given by the additional carbon dioxide. Higher CO_2 concentrations stimulate photosynthesis, enabling the plants to fix carbon at a higher rate. This is why in glasshouses additional CO_2 may be introduced artificially to increase productivity. The effect is particularly applicable to what are called C3 plants (such as wheat, rice and soya bean), but less so to C4 plants (for example maize, sorghum, sugar-cane, millet and many pasture and forage grasses). Under ideal conditions it can be a large effect (for doubled CO_2, up to 25 per cent for wheat and rice and up to 40 per cent for soybean)[38]. However, under real conditions on the large scale where water and nutrient availability are also important factors influencing plant growth, the increases, although difficult to estimate or measure, are likely to be substantially less than the ideal[39].

manipulation available today, there should be little difficulty in matching crops to new climatic conditions over large parts of the world. At least, that is the case for crops that mature over a year or two. Forests reach maturity over much longer periods, from decades up to a century or even more. The projected rate of climate change is such that, during this time, trees may find themselves in a climate to which they are far from suited. The temperature regime or the rainfall may be substantially changed, resulting in stunted growth or a greater susceptibility to disease and pests.

An example of adaptation to changing climate is the way in which farmers in Peru adjust the crops they grow depending on the climate forecast for the year[40]. Peru is a country whose climate is strongly influenced by the cycle of El Nino events described in chapters 1 and 5. Two of the primary crops grown in Peru, rice and cotton, are very sensitive to the amount and the timing of rainfall. Rice requires large amounts of water; cotton has deeper roots and is capable of yielding greater production during years of low rainfall. In 1983, following the 1982–83 El Nino event, agricultural

Climate change and world food supply

A detailed study of the effect of climate change on world food supply has been carried out at the Environmental Change unit at Oxford University with assistance from agricultural scientists in 18 countries[41]. For the details of climate change the results from several different models (of the kind described in Chapter 5) were used for the climate under doubled CO_2 concentration. Potential changes in national grain crop yields were estimated for wheat, rice, maize (which between them account for about 85 per cent of world cereal exports) and soybean (which accounts for about two-thirds of the trade in protein cake equivalent). These national crop yield changes were extrapolated to provide yield change estimates for other countries and other crops.

Various assumptions were made about the degree of adaptation by farmers. A world food trade model was employed to allow for possible levels of population growth and economic change.

On the assumption that the climate changes being estimated occur by the year 2060, assuming also a continuation in current trends in economic growth rates, partial trade liberalization and medium population growth rates, the main findings of the study are:

■ The negative effects of climate change are to some extent compensated for by increased productivity due to the fertilization effect of increased CO_2 (see box describing this) assumed in their model. With a modest level of farm-level adaptation (for example minor shifts in planting dates, changes in crop variety) the net effect of climate change is to reduce global cereal production from what it would be without climate change, by up to 5 per cent. This reduction can be overcome by more major forms of adaptation such as the installation of irrigation.

■ Climate change would increase the disparities in cereal production between developed and developing countries. In the model, production in the developed world tended to benefit from climate change (possible increase of around 5 per cent) whereas production in developing nations declined (by around 10 per cent) as a result of climate change. Adaptation at the farm level does little to reduce these disparities.

■ Cereal prices and thus the population at risk of hunger in developing countries are likely to increase despite adaptation.

The authors emphasize that, although the models and the methods they have employed are comparatively complex, there are many factors which have not been taken into account. For instance, they have not considered the availability of water supplies for irrigation. Further, (see Chapter 6) scientists are not yet very confident in the regional detail of climate change. The results, therefore, although giving a general indication of the changes that could occur, should not be treated as a detailed prediction. They highlight the importance of studies of this kind as a guide to future action.

production dropped by 14 per cent. By 1987 forecasts of the onset of El Nino events had become sufficiently good for Peruvian farmers to take them into account in their planning. In 1987, following the 1986–87 El Nino, production actually increased by 3 per cent, thanks to a useful forecast.

The availability of water is the most important of the factors affecting agriculture and food production. The vulnerability of water supplies to climate change carries over into a vulnerabilty in the growing of crops and the production of food. Thus the arid or semi-arid areas, mostly in developing countries, are most at risk. A further factor, which can actually lead to increased production as a result of climate change, is the boost to growth which is given, particularly to some crops, by increased atmospheric carbon dioxide (see box on p 103).

Detailed studies have been carried out of the sensitivity to climate change of the major crops which make up a large proportion of the world's food supply (for an example see box). They have used the results of models to estimate the changes in temperature and precipitation which might be expected when the atmosphere has a doubled carbon dioxide concentration. They have also included the possible effects of economic factors and the possibility of modest levels of adaptation. Such models are at an early stage of development and cannot be employed as detailed predictions. They tend to show that, with appropriate adaptation, the effect of the average climate change which would occur with doubled carbon dioxide on the supply of food for the

world as a whole is not likely to be very large. What has not been studied to any extent is the likely effect of climate extremes (especially of the incidence of drought) on global food supply. Nor have the studies considered the integrity of the world's soils, which are currently being lost by erosion at an alarming rate[42]. Further, a serious issue exposed by the studies is that climate change is likely to affect countries very differently. Production in developed countries may well increase, whereas that in many developing countries is likely to decline as a result of climate change. The disparity between nations will tend to become larger, as will the number of those at risk from hunger.

In looking to future needs, two activities which can be pursued now are particularly important. First, there is large room for technical advances in agriculture in developing countries where much is still very primitive. In particular, there needs to be continued development of programmes for crop breeding and management, especially in conditions of heat and drought. These can be immediately useful in the improvement of productivity in marginal environments today. Secondly, as was seen earlier when considering fresh water supplies, improvements need to be made in the availability and management of water for irrigation, especially in arid or semi-arid areas of the world.

Now to summarize the likely effect of global warming on agriculture and food supplies: although studies of the sensitivity of crop production and food supply are at an early stage, there is yet no

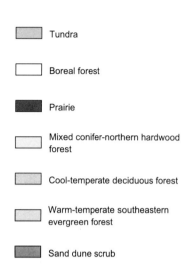

Tundra

Boreal forest

Prairie

Mixed conifer-northern hardwood forest

Cool-temperate deciduous forest

Warm-temperate southeastern evergreen forest

Sand dune scrub

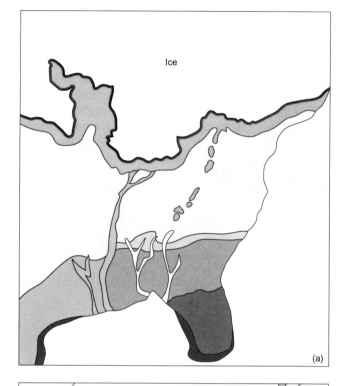

FIG. 7.9 Vegetation maps of the south-eastern United States during past climate regimes[43]: (a) for 18,000 years ago at the end of the last ice age, (b) for 10,000 years ago, (c) for 5,000 years ago when conditions were similar to the present; the vegetation map for 200 years ago is similar to that in (c).

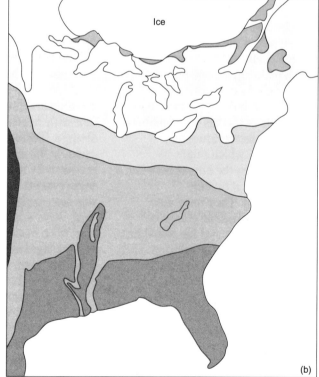

106

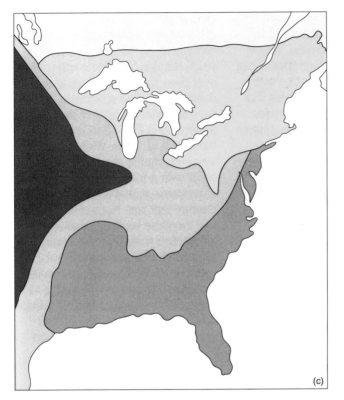

(c)

strong evidence that the effect of climate change on global food supply is likely to be large. What needs urgent research is how well world agriculture will respond to extremes, such as prolonged droughts. Careful studies of this should be carried out as soon as possible.

It is fairly clear, however, that an increasing mismatch is likely to develop between developed and developing countries. The surplus of food in developed countries is likely to increase, while in developing countries there will be large population increases coupled with a likely relative decrease in food production. Such a situation will raise enormous problems, especially in the developing world.

One of these will be that of employment. Agriculture is the main source of employment in developing countries; people need employment to be able to buy food. With changing climate, as some agricultural regions shift, people will tend to attempt to migrate to places where they might be employed in agriculture. With the pressures of rising populations such movement is likely to be increasingly difficult and we can expect large numbers of environmental refugees.

The impact on natural ecosystems

A little over 10 per cent of the world's land area is under cultivation. The rest is to a greater or lesser extent unmanaged by humans. Of this about 30 per cent is natural forest. The variety of plants and animals which constitute a local ecosystem is sensitive to the climate, the type of soil and the availability of water.

Ecologists divide the world into biomes—regions characterized by their distinctive vegetation. This is well illustrated by information about the distribution of vegetation over the world during past climates (Fig. 7.9), which indicates what species and what ecosystems are most likely to flourish under different climatic regimes.

The information gleaned from paleo sources could be used to produce maps of the optimum distribution of natural vegetation under the climate scenarios expected to occur with global warming, since climate is the dominant factor determining the distribution of biomes (Fig. 7.10)—changes in climate alter the suitability of a region for different species, and change the competitiveness of different species within an ecosystem, so that even relatively small changes in climate will lead, over time, to large changes in the composition of an ecosystem.

However, changes of the kind illustrated in Fig. 7.9 took place over thousands of years. With global warming similar changes in climate occur over a few decades. Most ecosystems cannot respond or migrate that fast. Natural ecosystems will therefore become increasingly unmatched to their environment. How much this matters will vary enormously from species to species; some are much more vulnerable to changes in average climate or climate extremes than others. But all will become more prone to disease and

FIG. 7.10 The pattern of world biome types related to mean annual temperature and precipitation. Other factors, especially the seasonal variations of these quantities, affect the detailed distribution patterns (after Gates[44]).

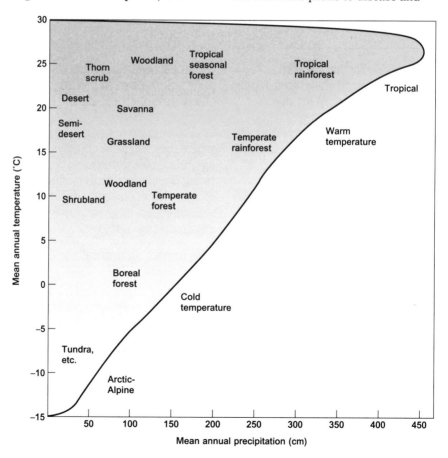

attack by pests. Any positive effect from added fertilization due to increased carbon dioxide is likely to be more than outweighed by negative effects from other factors.

Trees are long-lived and take a long time to reproduce. Because it is not easy for them to respond quickly to climate change, the world's forests are likely to be the most affected. The decline in the health of many forests in recent years has received much attention, especially in Europe and North America where much of it is thought to be due to acid rain and other pollution originating from heavy industry, power stations and motor cars. Not all damage to trees, however, is thought to have this origin. Studies in several regions of Canada, for instance, indicate that the dieback of trees there is related to changes in climatic conditions, especially to successions of warmer winters and drier summers[45]. In some cases it may be the double effect of pollution and climate stress causing the problem; trees already weakened by the effects of pollution fail to cope with climate stress when it comes. The assessment of the impact of climate change carried out for the MINK region of the United States (see box on p 100) concluded that, under the warmer, drier conditions of the analogue climate they studied, decline and dieback of the forested part of the region would reduce the mass of timber in the forest by 10 per cent over twenty years[46]. The results of these studies are indicative of the more serious levels of forest dieback which are likely to occur with the rapid rate of climate change expected with global warming.

A further concern about natural ecosystems, especially forests, relates to the diversity of species which they contain. Much concern is being expressed at the moment because of the threat to biodiversity from extensive deforestation. There is a danger that the rapid climate change expected as a result of global warming will also lead to significant loss of species in some ecosystems—although the likely effect is difficult to quantify.

So far we have been considering ecosystems on land. What about those in the oceans; how will they be affected by climate change? Although we know much less about ocean ecosystems, there is considerable evidence that biological activity in the oceans has varied during the cycle of ice ages. Chapter 3 noted (see box on p 32) the likelihood that it was these variations in marine biological activity which provided the main control on atmospheric carbon dioxide concentrations during the past million years (Fig. 4.4). The changes in ocean water temperature and the possible changes in some of the patterns of ocean circulation are likely to result in changes in the regions where upwelling occurs and where fish congregate. Some fisheries could collapse and others expand. At the moment the fishing industry is not well adapted to address major change[47].

The impact on human health

Human health is dependent on a good environment. Many of the factors that lead to a deteriorated environment also lead to poor health. Pollution of the atmosphere, polluted or inadequate water supplies, and poor soil (leading to poor crops and inadequate nutrition) all present dangers to human health and wellbeing and assist the spread of disease. As has been seen so far in considering the impacts of global warming, many of these factors will tend to be exacerbated through the climate change which will occur in the warmer world.

How about direct effects of the climate change itself on human health[48]? Humans can live comfortably in very varying conditions and have great ability to adapt to a wide range of climates. However, really extreme conditions are not very tolerable and can encourage the transmission of some diseases. The main difficulty in assessing the impact of climate change on health is that of unravelling the influences of climate from the large number of other factors (including other environmental factors) that affect health.

The main direct effect of climate change on humans themselves will be that of heat stress in the extreme high temperatures that will become more frequent and more widespread. Studies using data from large cities where heat waves commonly occur show death rates which can be doubled or tripled during days of unusually high temperatures[49]. Although such an episode may be followed by a period when fewer deaths occur, showing that some of the deaths would in any case have occurred about that time, most of the increased mortality seems to be directly associated with the excessive temperatures. It might be thought that compensation for the periods of excessive heat would be provided by fewer occasions of severe cold. Studies tend to show, however, that increased mortality due to periods of excessive heat will considerably exceed any reduction in the periods of cold, especially amongst old people who find it particularly difficult to cope with high temperatures[50].

A further likely impact of climate change on health is the increased spreading of diseases in a warmer world. Many insect carriers of disease thrive better in warmer and wetter conditions. For instance, epidemics of diseases such as viral encephalitides carried by mosquitoes are known to be associated with the unusually wet conditions which occur in the Australian, American and African continents associated with different phases of the El Nino cycle[51]. Some diseases, currently largely confined to tropical regions, with warmer conditions could spread into mid-latitudes. Malaria is an example of such a disease which is spread by mosquitoes under conditions which are optimum in the temperature range of 15–32°C with humidities of 50–60 per cent. Other diseases which are likely to spread for the same reason are yellow fever, dengue fever and lymphatic filariasis[52].

The arguments presented here suggest that the impact of climate change on health can be large. However, the factors involved in all cases are highly complex; any quantitative conclusions will require a great deal more careful study of the direct effects of climate on humans and of the epidemiology of the diseases which are likely to be particularly affected. An international project called Global Health Watch has been proposed for the collection of the data required for such studies[53].

Costing the impacts

It is clearly not an easy task to estimate the likely cost to the world community of the various impacts listed above. For a costing to be at all realistic, especially when it is to apply to periods of decades into the future, it must account not only for direct damage but also for the possibilities of adaptation. However, even crude attempts at costing help to provide an idea of the size of the problem.

The most detailed studies of impacts have been carried out for the United States and have been summarized by William Cline (see box on p 109). For those impacts against which some value of damage can be placed, the total figure is about 60 thousand million dollars per annum or about 1 per cent of the US Gross Domestic Product (GDP) in 1990. For other countries in the developed world, estimates of the cost of impacts in terms of percentage of GDP are similar; for the developing world, figures of annual cost of 2 per cent or more of GDP have been estimated.

As the authors of these economic studies explain, their estimates are crude, are based on very broad assumptions and should not be considered as precise values. They

Estimating the costs of global warming

Studies[54] of the costs of losses or damage for the United States arising from global warming under a business-as-usual scenario of greenhouse gas emissions have been analyzed by W.R. Cline[55], who emphasizes the large uncertainty associated with the estimates. The results have been presented in terms of average annual costs between now and the year 2050, when the amount of global warming due to the increase of greenhouse gases from their pre-industrial values is expected to be about 2.5°C. Table 7.1 (below) summarizes the largest contributions to the estimated costs set against the areas of impact presented above.

Estimates (in thousand millions of US dollars [1990 value] per year) of the cost of the impacts of climate change on the United States.

Loss of land and dyke construction due to sea-level rise	7
Loss of water supplies	7
Losses in the agriculture and forestry sectors	21
Increased morbidity	6
Increased electricity for air conditioning	10
Other factors	11
Total	62

This figure of 62 thousand million dollars is about 1 per cent of the GDP of the United States. Studies[54] of the costs arising from the likely impacts in the rest of the world come up with a similar figure in terms of percentage of GDP in developed countries but with higher figures—between about 2 per cent and 6 per cent of GDP—for developing countries, although again the authors of the studies emphasize the preliminary nature and uncertainty of the estimates.

provide us with a first indication of the scale of the problem in economic terms. Later chapters will compare them with the cost of taking action to slow the onset of global warming or reduce its overall magnitude.

An initial inspection of the figures in the table might suggest that the costs of global warming, although large, at a percent or two of Gross World Product (GWP) are not impossibly large, and that we could probably buy our way out of the problems of its impact. There are, however, as Cline points out, two factors which have been omitted from the studies cited above. The first is that the studies have only been concerned with the impacts of global warming up to about the middle of the next century. The longer-term impacts, if the growth of greenhouse gases continues along a business-as-usual scenario, are likely to be much greater. The second factor is that not all impacts can be quantified in terms of economic costs. The loss of human amenity, natural amenity or the loss of species cannot be easily expressed in money terms. This can be emphasized by focusing on those who are likely to be particularly disadvantaged by global warming. Most of them will be in the developing world at around the subsistence level. They will find their land is no longer able to sustain them because it has been lost either to sea-level rise or to extended drought. They will therefore wish to migrate and will become environmental refugees.

It has been estimated that, under a business-as-usual scenario, the total number of persons displaced by the impacts of global warming could total in the order of 150 million by the year 2050 (or about 3 million per year on average)—about 100 million due

to sea-level rise and coastal flooding and about 50 million due to the dislocation of agricultural production mainly due to the incidence and location of areas of drought[56]. The cost of resettling 3 million displaced person per year has been estimated at between $1,000 and $5,000 per person, giving a total of about $4 thousand million per year[57]. What the estimated cost for resettlement does not include, however, (as the authors of the study themselves emphasize) is the human cost associated with displacement. Nor does it include the social and political instabilities which ensue when substantial populations are seriously disrupted because their means of livelihood has disappeared. The effects of these could be very large indeed.

The overall impact of global warming

The incidence of various impacts of global warming is complex and far from uniform over the world.

■ There are many ways in which our current environment is being degraded due to human activities; global warming will tend to exacerbate these degradations. Sea-level rise will make the situation worse for low-lying land which is subsiding because of the withdrawal of groundwater and because the amount of sediment required to maintain the level of the land has been reduced. The loss of soil due to overuse of land or deforestation will be accelerated, with increasing droughts or floods in some areas. In other places, extensive deforestation will lead to drier climates and less sustainable agriculture.

■ It has been emphasized throughout that global warming will lead to changes in temperature and precipitation in many places. To respond to the impacts from these changes, there will be the need to adapt. In many cases this will involve changes in infrastructure, for instance new sea defences or water supplies. Many of the impacts of climate change will be adverse, but even when the impacts in the long term turn out to be beneficial, in the short term the process of adaptation will mostly have a negative impact and involve cost.

■ The most important impact is on water supplies, which are in any case becoming increasingly critical in many places. Some parts of the world are expected to become warmer and drier, especially in summer, with a greater likelihood of droughts; in other parts a greater incidence of floods is expected.

■ Through adaptation to different crops and practices, first indications are that the total of world food supplies can be maintained despite climate change. However, the disparity in food supplies between the developed and the developing world will almost certainly become larger.

■ Because of the likely rate of climate change, there will also be a serious impact on natural ecosystems, especially at mid to high latitudes; forests especially will be affected. In a warmer world longer periods of heat stress will have an effect on human health; warmer temperatures will also encourage the spread of certain tropical diseases, such as malaria, to higher latitudes.

■ Economists have atttempted to estimate the average annual cost in money terms of the impacts which would arise under the climate change likely for the IPCC business-as-usual scenario of greenhouse gas emissions. Averaged over the world these estimates are typically around 1 per cent of GWP, with the cost for developing countries being on average at least twice as large as that for developed countries. But they do not take into account the cost in human terms nor the substantial social and political disruption the impacts will bring. In particular, it is estimated that there could be up to 3 million new environmental refugees each year or over 150 million by the middle of next century.

It is important to bear in mind that these estimates have concentrated on the doubled carbon dioxide scenario (in other words, the next 50 or 60 years). Soon after the end of next century, under the IPCC business-as-usual scenario (in other words, if strong action is not taken to curb carbon dioxide emissions) a further doubling of the equivalent carbon dioxide concentration will have occurred and it will be continuing to rise. The impacts of the additional climate change which would occur with a second effective doubling of carbon dioxide are likely to be substantially more severe than those of the first doubling.

That, of course, is a lot further away in time; perhaps for that reason it has not been given much attention. However, because of the long life-time of some greenhouse gases, because of the long memory of the climate system, because some of the impacts may turn out to be irreversible and also because of the time taken for human activities to respond and change course, it is important to have an eye on the longer term. The much more severe impacts which can be expected at longer time horizons increase the imperative now to take the necessary action.

However, many will ask why we should be concerned about the state of the Earth so far ahead in the future. Can we not leave it to be looked after by future generations? The next chapter will give something of my personal motivation for caring about what happens to the Earth in the future as well as now.

FOOTNOTES

1 See for instance A. Goudie, *The human impact on the natural environment*, MIT Press, 1986.

2 J. Oerlemans and J.P.F. Fortuin, 'Sensitivity of glaciers and small ice-caps to greenhouse warming', *Science*, **258**, 1992, pp 115–17.

3 R.Warrick and J.Oerlemans 'Sea-level rise' in *Climate Change, the IPCC Scientific Assessment*, eds J.T. Houghton, G.J. Jenkins and J.J. Ephraums, CUP, 1990 and *Climate Change 1992, the Supplementary Report to the IPCC Scientific Assessment*, eds J.T. Houghton, B.A. Callander and S.K. Varney, CUP, 1992. The curves in the IPCC 1990 report have been modified to apply to the IS 92a scenario.

4 See for instance C.J. van der Veen, 'State of balance of the cryosphere', *Rev. Geophys.*, **29**, 1991, pp 433–55.

5 From Warrick and Oerlemans 1990 *loc. cit.*

6 J.M. Broadus, 'Possible impacts of, and adjustments to, sea-level rise: the cases of Bangladesh and Egypt' in *Climate and sea-level change: observations, projections and implications*, eds R.A. Warrick, E.M. Barrow and T.M.L. Wigley, CUP, 1993, pp 263–75. Note that, because of variations in the ocean structure, sea level rise would not be the same everywhere. In Bangladesh it would be somewhat above average (see J.M. Gregory 'Sea-level changes under increasing CO_2 in a transient coupled ocean-atmosphere experiment,' *J. Climate*, 6, 1993, pp 2247–62).

7 From J.M. Broadus, 1993, *loc. cit.*—adapted from J.D.Milliman *et al.* 'Environmental and economic implications of rising sea-level and subsiding deltas: the Nile and Bangladesh examples', *Ambio*, **18**, 1989, pp 340–45.

8 From a report prepared in 1989 by F.U. Muhtab, *Effect of climate change and sea level rise on Bangladesh*, Commonwealth Secretariat, London.

9 J.M. Broadus *loc. cit.*

10 J.D. Milliman *loc. cit.*

11 From report entitled *Climate change due to the greenhouse effect and its implications for China* published in 1992 by the Worldwide Fund for Nature, Gland, Switzerland.

12 J.W. Day *et al.*, 'Impacts of sea-level rise on coastal systems with special emphasis on the Mississippi river deltaic plain' in *Climate and sea-level change: observations, projections and implications*, eds R.A. Warrick, E.M. Barrow and T.M.L. Wigley, CUP, 1993, pp 276–296.

13 K.M. Clayton 'Adjustment to greenhouse gas induced sea-level rise on the Norfolk coast—a case study' in *Climate and sea-level change: observations, projections and implications*, eds R.A. Warrick, E.M. Barrow and T.M.L. Wigley, CUP, 1993, pp 310–21.

14 J.G. de Ronde, 'What will happen to the Netherlands if sea-level rise accelerates?' in *Climate and sea-level change: observations, projections and implications*, eds R.A. Warrick, E.M. Barrow and T.M.L. Wigley, CUP, 1993, pp 322–35.

15 More details of the impacts on small islands and other impacts of sea-level rise in *Climate Change: the IPCC Impacts Assessment*, eds W.J.McG. Tegart, G.W. Sheldon and D.C. Griffiths, Australian Government Publishing Service, Canberra, 1990, Chapter 6.

16 *Climate Change: the IPCC Impacts Assessment*, Chapter 6 and *The potential socio-economic effects of climate change in south-east Asia*, UNEP Report 1992.

17 From *The world environment 1972–1992*, eds M.K.Tolba & O.A.El-Kholy, Chapman and Hall, 1992, p 84.

18 From J.W.Maurits la Riviere, 'Threats to the world's water', *Sci Amer.* **261**, September 1989, pp 48–55.

19 From *The world environment 1972–1992, op. cit.* p 85—adapted from Shiklomanov (1988).

20 P.H. Gleick 'Regional hydrologic consequences of increases in atmospheric CO_2 and other trace gases', *Climatic Change*, **10**, 1987, pp 137–61.

21 Quoted by Geoffrey Lean in 'Troubled Waters', the colour supplement to the *Observer* newspaper, 4 July 1993.

22 J.A. Dracup and D.R. Kendall, 'Floods and droughts' in *Climate change and US water resources*, ed. P.E. Waggoner, Wiley, 1990, pp 243–67.

23 P.H. Gleick, 'Vulnerability of water systems' in *Climate change and US water resources*, ed. P.E. Waggoner, Wiley, 1990, pp 223–40.

24 'Towards an integrated impact assessment of climate change: the MINK study', ed. N.J. Rosenberg, *Climatic Change*, **24**, Nos 1–2, 1993, Kluwer Academic Publishers.

25 The MINK study on water resources is described by K.D. Frederick, *Climatic Change*, **24**, 1993, pp 83–115.

26 From *The world environment 1972–1992*, eds M.K. Tolba & O.A. El-Kholy, Chapman and Hall, 1992, p 134.

27 From *The world environment 1972–1992*, eds M.K. Tolba & O.A. El-Kholy, Chapman and Hall, 1992, p 135.

28 J. Lean and P.R.Rowntree, 'A simulation of the impact of Amazonian deforestation on climate using an improved canopy representation' *Q.J.R.Meteorol.Soc.*, **119**, 1993, pp 509–530. Similar results with somewhat larger reductions in rainfall have been reported by A.Henderson-Sellers *et al.*, 'Tropical Deforestation: Modelling local to regional scale climate change', *J. Geophys. Res.*, **98**, 1993, pp 7289–315.

29 M.F. Mylne and P.R. Rowntree, 'Modelling the effects of albedo change associated with tropical deforestation', *Climatic Change*, **21**, 1992, pp 317–43.

30 Quoted in J.F.B. Mitchell *et al.*, 'Equilibrium climate change and its implications for the future' in *Climate Change, the IPCC Scientific Assessment*, eds J.T. Houghton, G.J. Jenkins and J.J. Ephraums, CUP, 1990, pp 131–72.

31 See for instance J.W. Maurits de la Riviere *loc. cit.*

32 *Climate Change and US Water Resources*, ed. P.E. Waggoner, Wiley, 1992.

33 M. Lal, 'Water resources of the south-east Asian region in a warmer atmosphere', *Advances in Atmospheric Sciences*, **11**, **No 2**, 1994, Academia Sinica, China.

34 M. Lal, 'Global climate change and its impact on Indian agriculture' in a state-of-the-art publication of ICAR, New Delhi, entitled *Global climate change and Indian agriculture*, 1994.

35 M. Lal, 'Water resources of the south-east Asian region in a warmer atmosphere', *Advances in Atmospheric Sciences*, **11**, No 2, 1994, Academia Sinica, China.

36 *Climate Change: the IPCC Impacts Assessment*, Chapter 6, p 4.24.

37 P.R. Crosson and N.J. Rosenberg, 'Strategies for agriculture', *Scien. Amer.*, September 1989, pp 78–85.

38 C. Rosenzweig, M.L. Parry, G. Fischer & K. Frohberg, *Climate Change and World Food Supply*, Research Report **No 3**, 1993, Environmental Change Unit, University of Oxford.

39 For an estimate of the global effect of CO_2 fertilization on ecosystems see J.M. Melillo *et al.*, 'Global climate change and terrestrial net primary production', *Nature*, 1993, pp 234–40.

40 Information in proposal, edited by A.D. Moura, for an International Research Institute for Climate Prediction report prepared in 1992 for the International Board for the TOGA project, World Meteorological Organization, Geneva.

41 C. Rosenzweig *et al.* 1993, *loc. cit.*

42 *The world environment 1972–1992, op. cit.*

43 Adapted from D.M. Gates , *Climate Change and its Biological Consequences*, Sinauer Associates Inc. Sunderland, Mass., USA 1993, p63.; the original source is Delcourt and Delcourt 1981. This book contains a detailed review of natural ecosystems and climate change.

44 D.M. Gates, 1993 *loc. cit.*

45 D.M. Gates, 1993 *op. cit.* p 77.

46 M.D. Bowes and R.A. Sedjo, 'Impacts and responses to climate change in forests of the MINK region', *Climatic Change*, **24**, 1993, pp 63–82.

47 *Climate Change: the IPCC Impacts Assessment*, 1990, pp 6–20.

48 A series of weekly articles under the general heading of 'Climate and Health' in *The Lancet* from 23 October to 11 December 1993 has addressed many of the issues.

49 I.S. Kalkstein, 'Direct impact in cities', *The Lancet*, **342**, 1993, pp 1397–1399.

50 Studies carried out by the US Environmental Protection Agency, quoted by W.R. Cline, *The economics of global warming*, Institute for International Economics, Washington DC 1992, pp 116–17.

51 N. Nicholls, 'El Nino-Southern Oscillation and vector-borne disease', *The Lancet*, **342**, 1993, pp 1284–85. The El Nino cycle is described above in Chapter 5.

52 W.H. Weihe and R. Mertens, 'Human well-being, diseases and climate', in *Climate Change: Science, Impacts and Policy*, eds J. Jager and H.L. Ferguson, CUP, 1991, pp 345–59; G.C. Cook, 'Effect of global warming on the distribution of parasitic and other infectious diseases: a review', *J. Roy. Soc. Med.*, **85**, 1992, pp 688–91.

53 A. Haines *et al.*, 'Global Health Watch: monitoring impacts of environmental change', *The Lancet*, **342**, 1993, pp 1464–69.

54 W.R. Cline, *Economics of global warming*, Institute for International Economics, Washington DC USA 1992, also see N. Adger and S. Fankhauser, 'Economic analysis of the greenhouse effect: optimal abatement level and strategies for mitigation', *Int. J. of Environment and Pollution*, **3**, 1993, pp 104–19.

55 W.R. Cline, *op. cit.*

56 N.Myers, 'Environmental refugees: how many ahead?' submitted to *Bioscience*, 1993; also Adger and Fankhauser *loc. cit.*

57 N. Adger and S. Fankhauser, *loc cit.*

8 *Why Should We Be Concerned?*

I *have been describing the likely changes in climate which may occur as a result of human activities, and the impact these may have in different parts of the world. But large and potentially devastating changes are likely to be a generation or more away. So why should we be concerned? What responsibility, if any, do we have for the planet as a whole and the great variety of other forms of life which inhabit it? And does our scientific knowledge in any way match up with other insights, for instance religious ones, regarding our relationship with our environment? In this chapter I want to digress from the detailed consideration of global warming (to which I shall return) in order briefly to explore these very fundamental questions and to present something of my personal viewpoint on them.*

Earth in the Balance

Al Gore, the Vice-President of the United States, entitled his recent book on the environment *Earth in the Balance*[1], implying that there are balances in the environment which need to be maintained. A small area of a tropical forest possesses an ecosystem which contains some thousands of plant and animal species, each thriving in its own ecological niche in close balance with the others. Balances are also important for larger regions and for the Earth as a whole. These balances can be highly precarious, especially where humans are concerned. One of the first to point this out was

Rachel Carson in her book *Silent Spring*[2], first published in 1962, which described the damaging effects of pesticides on the environment. We are an important part of the global ecosystem; as the size and scale of our activities continue to escalate, so can the seriousness of the disturbances we cause to the overall balances of nature. Some examples of this were given in the last chapter.

It is important that we recognize these balances, in particular the careful relationship between humans and the world around us. It needs to be a balanced and harmonious relationship in which each generation of humans should leave the Earth in a better state, or at least in as good a state as they found it. The word that is often used for this is sustainability—politicians talk of sustainable development. This principle, and its link with the harmonious relationship between humans and nature, was given prominent place by the United Nations Conference on Environment and Development held at Rio de Janeiro in June 1992. The first principle in a list of 27 at the Rio Declaration adopted by the Conference is 'Human beings are at the centre of concerns for sustainable development. They are entitled to a healthy and productive life in harmony with nature.'

However, despite such statements of principle from a body such as the United Nations, many of the attitudes which we commonly have to the Earth are, however, neither balanced, harmonious nor sustainable. Some of these are briefly outlined in the following paragraphs.

Exploitation

Humankind has over many centuries been exploiting the Earth and its resources. It was at the beginning of the Industrial Revolution some two hundred years ago that the potential of the Earth's minerals began to be realized. Coal, the result of the decay of primaeval forests and laid down over many millions of years, was the main source of energy for the new industrial developments. Iron ore to make steel was mined in vastly increased quantities. The search for other metals such as zinc, copper and lead was intensified until today many millions of tons are mined each year. Around 1960, oil took over from coal as the dominant world source of energy; oil and gas between them now supply over

twice the energy supplied by coal.

We have not only been exploiting the Earth's mineral resources. The Earth's biological resources have also been attacked. Forests have been cut down on a large scale to make room for agriculture and for human habitation. Tropical forests are a particularly valuable resource, important for the maintenance of the climate of tropical regions. They have also been estimated to contain perhaps half of all the Earth's biological species. Yet only about half of the mature tropical forests which existed a few hundred years ago still stand[3]. At the present rate of destruction virtually all will be gone in less than fifty years.

Much of this exploitation has been carried out with little or no thought as to whether this use of natural resources is a responsible one. Early in the Industrial Revolution it seemed that resources were essentially limitless. Later on, as one source ran out others became available to more than take its place. Even now, for most minerals new sources are being found faster than present sources are being used. But the growth of use is such that this situation cannot continue. In many cases known reserves or even likely reserves will begin to run out during the next hundred or few hundred years. These resources have been laid down over many millions if not thousand millions of years. Nature took about a million years to lay down the amount of fossil fuel that we now burn worldwide every year—and in doing so it seems that we are causing rapid change of the Earth's climate. Such a level of exploitation is clearly not in balance, not harmonious and not sustainable.

Back to nature

Almost the reverse of this attitude is the suggestion that we all adopt a much more primitive lifestyle and give up a large part of industry and intensive farming—that we effectively put the clock back two or three hundred years to before the Industrial Revolution. That sounds very seductive and some individuals can clearly begin to live that way. But there are two main problems.

The first is that it is just not practical. The world population is now some six times what it was two hundred years ago and about three times that of fifty years ago. The world cannot be adequately fed without intensive farming and without modern methods of food distribution. Further, most people in the developed world just would not be prepared to be without the technical aids—electricity, central heating, refrigerator, washing machine, television and so on—which give the freedom, the interest and the entertainment which is so much taken for granted. In fact, people in the developing world are also beginning to take advantage of and enjoy these aids to a life of less drudgery and more freedom.

The second problem is that it fails to take account of human creativity. Human scientific and technical development cannot be frozen at a given point in history, insisting that no further ideas can be developed. A proper balance between humans and the environment must leave room for humans to exercise their creative skills.

Again, therefore, a 'back to nature' viewpoint is neither balanced nor sustainable.

The technical fix

A third common attitude to the Earth is to invoke the 'technical fix'. As a senior environmental official from the United States said to me some years ago, 'We cannot change our lifestyle because of the possibility of climate change, we just need to fix the biosphere.' It was not clear just what he supposed the technical fixes would turn out to be. The point that he was making is that, in the past, humans have been so effective at developing new technology to meet the problems as they arise, can it not be assumed that this will continue? Concern about the future then turns into finding the 'fixes' as they are required.

On the surface the 'technical fix' route may sound a good way to proceed; it demands little effort and no foresight. It implies that damage can be corrected when it has been created rather than avoided in the first place. But damage already done to the environment by human activities is causing problems now. It is as if in looking after my home I decided not to carry out any routine maintenance but 'fixed' the failures as they occurred. For my home that would be a high risk route to follow: failure to rewire when necessary could easily lead to a disastrous fire. A similar attitude to the Earth is both arrogant and irresponsible. It fails to recognize the vulnerability of nature to the large changes which human activities are now able to generate.

Again, the 'technical fix' approach is neither balanced nor sustainable.

Future generations

Having described attitudes which are not balanced or harmonious in their relationship to the Earth and which fail to contribute to sustainability, I shall now turn to describe attitudes to the environment which are more acceptable in terms of the criteria I have set.

First, there is our responsibility to future generations. It is a basic instinct that we wish to see our children and our grandchildren well set up in the world and wish to pass on to them some of our most treasured possessions. A similar desire would be that they inherit from us an Earth which has been well looked after and which does not present them with more difficult problems than those we have had to face. But such an attitude is not universally held. I remember well, after the presentation I made on global warming to the British Cabinet at number 10, Downing Street, a senior politician saying to me that the problem would not become serious in his lifetime and could be left for its solution to the next generation. I do not think he had appreciated that the longer we delay in taking action, the larger the problem becomes and the more difficult to solve. We really do need to face up to the problem now for the sake of the next and subsequent generations. We have no right to act as if there is no tomorrow. We also have a responsibility to give to those who follow us a pattern for their future based on the principle of sustainable development.

The unity of the Earth

A second point of view sees us as having some responsibility, not just for all generations of humanity, but also for the larger world of all living things. We are, after all, part of that larger world. There is good scientific justification for this. We are

becoming increasingly aware of our dependence on the rest of nature and of the interdependences which exist between different forms of life, between living systems and the physical and chemical environment which surrounds life on the Earth— and indeed between ourselves and the rest of the universe.

The scientific theory named Gaia after the Greek Earth goddess and publicized particularly by James Lovelock emphasizes these interdependencies. Lovelock[4] points out that the chemical composition of the Earth's atmosphere is very different from that of our nearest planetary neighbours, Mars and Venus. Their atmospheres, apart from some water vapour, are almost pure carbon dioxide. The Earth's atmosphere, by contrast, is 78 per cent nitrogen, 21 per cent oxygen and only 0.03 per cent carbon dioxide. This composition is very far from chemical equilibrium and, so far as the major constituents are concerned, has remained substantially unchanged over many millions of years.

This very different atmosphere on the Earth has come about because of the emergence of life. Early in the history of life, plants appeared which photosynthesize, taking in carbon dioxide and giving out oxygen. There followed other living systems which 'breathe', taking in oxygen and giving out carbon dioxide. The presence of life therefore influences the environment, and living systems in turn adapt to the environment. It is the close match of the environment to the needs of life and its development which seems so remarkable and which Lovelock has emphasized. He gives many examples; I will quote one concerned with oxygen in the atmosphere. There is a critical

Daisyworld and life on the early Earth

Daisyworld is an imaginary planet spinning on its axis and orbiting a sun rather like our own. Only daisies live in Daisyworld; they are of two hues, black and white. The daisies are sensitive to temperature. They grow best at 20°C, below 5°C they will not grow and above 40°C they wilt and die. The daisies influence their own temperature by the way they absorb and emit radiation; black ones absorb more sunlight and therefore keep warmer than white ones.

In the early period of Daisyworld's history (Fig. 8.1), the sun is relatively cool and the black daisies are favoured because, by absorbing sunlight, they can keep their temperature closest to 20°C. Most of their white cousins die because they reflect sunlight and fail to keep above the critical 5°C. However, later in the planet's history, the sun becomes hotter. Now the white daisies can also flourish; both sorts of daisies are present in abundance. Later still as the sun becomes even hotter the white daisies become dominant as conditions become too warm for the black ones. Eventually, if the sun continues to increase its temperature even the white ones cannot keep below the critical 40°C and all the daisies die.

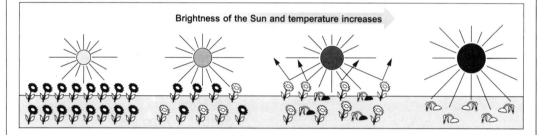

Brightness of the Sun and temperature increases

Daisyworld is a simple model employed by Lovelock[5] to illustrate the sort of feedbacks and self-regulation which occur in very much more complex forms within the living systems on the Earth.

Lovelock proposes a similar simple model as a possible description of the early history of life on the Earth (Fig. 8.2). The dashed line shows the temperature which would be expected on a planet possessing no life and with an atmosphere consisting, like our present atmosphere, mostly of nitrogen with about 10 per cent carbon dioxide. The rise in temperature occurs because the sun gradually became hotter during this period. About 3,500 million years ago primitive life appeared. Lovelock, in this model, assumes just two forms of life, bacteria which are anaerobic photosynthesizers—using carbon dioxide to build up their bodies but not giving out oxygen—and bacteria which are decomposers, converting organic matter back to carbon dioxide and

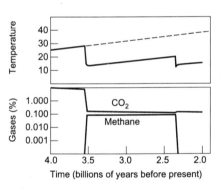

FIG. 8.1 **Daisyworld.**

FIG. 8.2. **Model of the Earth's early history, as proposed by Lovelock[6].**

methane. As life appears the temperature falls along with the concentration of the greenhouse gas, carbon dioxide. At the end of the period about 2,300 million years ago, more complicated life appears; there is an excess of free oxygen and the methane abundance falls to low values, leading to another fall in temperature, methane also being a greenhouse gas. The overall influence of these biological processes has been to maintain a stable and tolerable temperature for the Earth during this period.

:onnection between the oxygen concentration and the frequency of forest fires[7]. Below an oxygen concentration of 15 per cent, fires cannot be started even in dry twigs. At concentrations above 25 per cent fires burn extremely fiercely even in the damp wood of a tropical rain forest. Some species are dependent on fires for their survival; for instance, some conifers require the heat of fire to release their seeds from the seed pods. Above 25 per cent concentration of oxygen there would be no forests; below 15 per cent, the regeneration that fires provide in the world's forests would be absent. The oxygen concentration of 21 per cent is ideal.

It is this sort of connection that has driven Lovelock to propose that there is tight coupling between the organisms that make up the world of living systems and their environment. He has suggested a simple model of an imaginary world called Daisyworld (see box) which illustrates the type of feedback mechanisms which can lead to tight coupling. This model is similar to the one he has proposed for the biological and chemical history of the Earth during the first 1,000 million years after primitive life first appeared on the Earth some 3,500 million years ago.

The real world is, of course, enormously more complex than Daisyworld, which is why the Gaia hypothesis has led to so much debate. Lovelock's first statement in 1972 of the hypothesis[8] was that 'Life, or the biosphere, regulates or maintains climate and the atmospheric composition at an optimum for itself.' In his later writings he has introduced the analogy between the Earth and a living organism, introducing a new science which he

calls geophysiology[9]. In fact, his latest book is entitled *Gaia, the practical science of planetary medicine*[10].

An advanced organism such as a human being has many built-in mechanisms for controlling the interactions between different parts of the organism and for self-regulation. In a similar way, Lovelock argues, the ecosystems on the Earth are so tightly coupled to their physical and chemical environments that the ecosystems and their environment could be considered as one organism with an integrated 'physiology'. In this sense he believes that the Earth is 'alive'.

That elaborate feedback mechanisms exist in nature for control and for adaptation to the environment is not in dispute. But many scientists feel that Lovelock has gone too far in suggesting that ecosystems and their environment can be considered as a single organism. Although Gaia has stimulated much scientific comment it remains a hypothesis. What the debate has done, however, is to emphasize the interdependencies which connect all living systems to their environment.

There is the hint of a suggestion in the Gaia hypothesis that the Earth's feedbacks and self-regulation are so strong that we humans need not be concerned about the pollution we produce—Gaia has enough control to take care of anything we might do. Such a view fails to recognize the effect on the Earth's system of substantial disturbances, in particular the vulnerability of the environment with respect to its suitability for humans. To quote Lovelock[11], 'Gaia, as I see her, is no doting mother tolerant of misdemeanours, nor is she some fragile and delicate damsel in danger from brutal mankind. She is stern and tough, always keeping the world warm and comfortable for those who obey the rules, but ruthless in her destruction of those who transgress. Her unconscious goal is a planet fit for life. If humans stand in the way of this, we shall be eliminated with as little pity as would be shown by the micro-brain of an intercontinental ballistic nuclear missile in full flight to its target.'

Gaia is a scientific theory. But some have been quick to see it as a religious idea, supporting ancient religious beliefs. Many of the world's religions have drawn attention to the close relationship between humans and the Earth.

The Native American tribes of North America lived close to the Earth. One of their chiefs when asked to sell his land expressed his dismay at the idea and said[12], 'The Earth does not belong to man, man belongs to the Earth. All things are connected like the blood that unites us all.'

There is an ancient Hindu saying[13], 'The Earth is our mother, and we are all her children.' Modern civilization has taken us away from this feeling of closeness to the Earth. The Gaia scientific hypothesis can help to bring us back to recognize two things, first the inherent value of all parts of nature and secondly our dependence, as human beings, on the Earth and on our environment.

The Islamic faith teaches the value of sustainable development, for instance in a saying of the prophet Mohammed: 'He who revives a dead land will be rewarded accordingly, and that which is eaten by birds, insects and animals out of that land will be charity provided by God'. This underlines both our duty to care for the natural environment and our obligation to allow all living creatures their rightful place within it[14].

In looking for themes which emphasize the unity between humans and their environment, we need not confine ourselves to the Earth. There is a very much larger sphere in which a similar perspective of unity is becoming apparent. Some astronomers and cosmologists, overwhelmed by the size, scale, complexity, intricacy and precision of the universe, have begun to realize that their quest for an understanding of the evolution of the universe right from the 'Big Bang' some fifteen thousand million years ago is not just a scientific project but a search for meaning[15]. Why else has Stephen Hawking's book *A Brief History of Time*[16], in selling over six million copies, become one of the bestsellers of our time?

In this new search for meaning, the perspective has arisen that the universe was made with humans in mind—an idea expressed in some formulations of the anthropic principle[17]. Two particular pointers emphasize this. First, we have already seen that the Earth itself is fitted in a remarkable way for advanced forms of life. Cosmology is telling us that, in order for life on our planet to be possible, the universe itself at the time of the Big Bang and in its early history needed to be 'fine-tuned' to an incredible degree[18]. Secondly, there is the remarkable fact that human minds, themselves dependent on the whole universe for their existence, are able to appreciate and understand to some extent the fundamental mathematical structure of the universe's design[19]. As Albert Einstein commented, 'The most incomprehensible thing about the universe is that it is comprehensible.' In the theory of Gaia, the Earth itself is central and humans are dispensable; the insights of cosmology suggest that humans have a particular place in the whole scheme of things.

This section has recognized the intrinsic unity and interdependencies which exist not only on our Earth but also within the whole universe, and the particular place that we humans have in the universe. Being aware of these has large implications for our attitude to our environment.

Stewards of the Earth

The relationship between humans and the Earth which I have been advocating is often described as one of *stewardship*. We are on the Earth as its stewards. The word implies that we are carrying out our duty as stewards on behalf of someone else— but whom? Some environmentalists see no need to answer the question specifically, others might say we are stewards on behalf of future generations or on behalf of a generalised humanity. But a religious person would want to be more specific and say that we are stewards on behalf of God. I am a Christian myself and, in the next few paragraphs, want to explore the additional insights which are provided by a religious (and in particular by a Christian) view[20].

It is commonly thought that religious belief is not consistent with a scientific outlook. Scientists are, after all, looking for a description of the world in scientific terms which, when they know enough, they believe can be complete. For instance, scientists are looking for mechanisms to describe the 'fine-tuning' of the universe (these are known as 'Theories of Everything'!)

mentioned earlier. They are also looking for mechanisms to describe the interdependencies between living systems and the environment.

Suppose that scientists are successful in finding the mechanisms for which they are looking; will that negate the religious view that it is God who is responsible for the design of the universe and who had humans in mind in his creation? Not at all. What scientists are doing is trying to find out *how* the universe works; they cannot answer the ultimate questions as to *why* it is there. In the words of Johannes Kepler, who in the sixteenth century was one of the first of the modern observational astronomers, scientists are 'thinking God's thoughts after him'. The science we study is God's science; the mechanisms we discover are God's mechanisms. The fact that God may have built in automatic means to accomplish his purpose provides us with an even greater idea of what God is like. I would argue that scientific insights help me even more to believe that it is God's Earth, and that humans were created in God's image[21]!

However, not all religious people are so convinced that the Earth belongs to God. Reacting against the materialism of our time, they believe that God is only really interested in what is called the spiritual as opposed to the material part of life. Although their day to day living is dominated by the material, when it comes to the religious part of their lives, they feel the material has little or no part.

Some Christians home in on those verses in the Bible which concentrate on the transitory nature of material things. They see their future in heaven as a spiritual one and are not, therefore, concerned about the future of the Earth. Such an emphasis fails to do justice to the overall message of the Bible, which has a great deal to say about God's concern for the material creation and for its future.

Science and religion need to be seen as complementary ways of looking at truth, a point made strongly by Al Gore in *Earth in the Balance*[22] which lucidly discusses current environmental issues such as global warming. He blames much of our lack of understanding of the environment on the modern approach which tends to separate scientific study from religious and ethical issues. Science and technology are often pursued with a clinical detachment and without thinking about the ethical consequences. 'The new power derived from scientific knowledge could be used to dominate nature with moral impunity[23],' he writes. He goes on to describe the modern technocrat as 'this barren spirit, precinct of the disembodied intellect, which knows the way things work but not the way they are'[24]. However, he also points out[25] that 'there is now a powerful impulse in some parts of the scientific community to heal the breach' between science and religion. In particular, as we pursue an understanding of the Earth's environment, it is essential that scientific studies and technological inventions are not divorced from their ethical and religious context.

Gardeners of the Earth

A helpful picture of stewardship is found in the Judaeo-Christian tradition in the story of creation in the early chapters of the Bible. God's purpose, we are told, in making people was to care for the rest of creation—the idea of human stewardship of creation is a very old one! Adam and Eve were placed in a garden, the garden of Eden, 'to work it and take care of it'[26]. The animals, birds and other living creatures were brought to Adam in the garden for him to name them[27]. We are left with a picture of our first parents as 'gardeners' of the Earth; the Earth as God's garden and us as gardeners. What does our work as 'gardeners' imply? I want to suggest four things.

■ A garden provides food and water and other materials to sustain life and human industry. It is interesting that the Genesis story not only mentions food and water from the garden but also mineral resources. Regarding part of the garden we are told 'the gold of that land is good; aromatic resin and onyx are also there'[28]. The Earth provides resources for us humans to use as we need them.

■ A garden is to be maintained as a place of beauty. The trees in the garden of Eden were 'pleasing to the eye'[29].We are to live in harmony with the rest of creation and to appreciate the value of all parts of creation. Indeed, a garden is a place where care is taken to preserve the multiplicity of species, in particular those that are most vulnerable. Millions of people each year visit gardens which have been especially designed to show off the incredible variety and beauty of nature. Gardens are meant to be enjoyed.

■ A garden is a place where humans can be creative. Its resources provide for great potential. The variety of species and landscape can be employed to increase the garden's beauty and its productivity. Humans have learnt to generate new plant varieties in abundance and to use their scientific and technological knowledge coupled with the enormous variety of the Earth's resources to create new possibilities for life and its enjoyment. However, the potential of this creativity is such that increasingly we need to be aware of where it can take us; it has potential for evil as well as for good.

■ A garden is to be kept so as to be of benefit to future generations. In this context, I shall always remember Gordon Dobson, a distinguished scientist, who in the 1920s developed new means for the measurement of ozone in the atmosphere. His home outside Oxford in England possessed a large garden with many fruit trees. When he was 85, a year or so before he died, I remember finding him hard at work in his garden replacing a number of apple trees; in doing so he clearly had future generations in mind.

How well do we humans match up to the description of ourselves as gardeners caring for the Earth? Not very well, it must be said; we are more often exploiters and spoilers than cultivators. Some have placed part of the blame for this on attitudes[30] which they believe originate in the early chapters of Genesis, which talk of human beings having rule over creation and subduing it[31]. Those words, however, are not a mandate for unrestrained exploitation. The Genesis chapters also insist that human rule over creation is to be exercised under God, the ultimate ruler of creation, and with the sort of care exemplified by the picture of humans as 'gardeners'. Passages in the rest of the Bible

which are concerned with the care of land and with the environment emphasize the same requirement for care and stewardship[32].

Partnership with God

Many of the principles I have been enunciating are included at least implicitly in the declarations, conventions and resolutions which came out of the United Nations Conference on Environment and Development held in Rio de Janeiro in June 1992; indeed, they form the background of many statements emanating from the United Nations or from official national sources. We are not short of statements of ideals. What tend to be lacking are the capability and resolve to carry them out.

We are only too aware of the strong temptations we experience at both the personal and the national levels to use the world's resources to gratify our selfishness and greed. As we realize how formidable is the task of stewardship of the Earth, we may often feel that it is beyond the capability of the human race to tackle it adequately. But an important religious message is that we do not have to carry this responsibility on our own. Our partner is none other than God himself. The Genesis stories contain a beautiful description of this partnership where they speak of God 'walking in the garden in the cool of the day'[33].

Tragedy comes in that same chapter[34] where we are vividly told that humans disobeyed God and broke the partnership. The disasters we find everywhere in our environment speak eloquently of the consequences of that broken relationship.

But that did not put an end to God's purpose for humans or for the

Earth[35]. God continually offers a way back to that position of partnership[36]. One of the best known stories in the Old Testament is the story of Joseph[37], who interpreted the dream of Pharaoh, the ruler of Egypt, in terms of a climate forecast. There were to be seven years of plenty followed by seven when, because of drought, there would be poor crops. Pharaoh asked Joseph to manage the country's resources so that provision for the years of famine could be made during the years of plenty. Joseph saw the trying events in his life up to that time as part of God's plan to prepare him for the position of great responsibility which was now thrust upon him. He knew that his insight to interpret Pharaoh's dream and the ability for his new political and management task could only come from his close association with God.

Christianity teaches that this partnership between God and humanity reached its climax when in Jesus God became human, as the New Testament describes, thus demonstrating to the fullest possible extent God's commitment to the material world. When Jesus taught people about God, many of his examples came from human interaction with the natural world—sowing seed, looking after sheep, fishing. And in his parable of the talents[38], Jesus emphasized the importance of taking responsibility to the greatest possible extent; his harshest words were reserved for the servant who failed to use wisely the money entrusted to him by his master, but hid it in the ground.

In the Christian message the material and the spiritual are closely linked together. Jesus once said to his disciples, 'Without me you can do nothing[39].' This is often interpreted as relating particularly to the spiritual sphere and to religious activity. But I believe that Jesus meant it to be a much more comprehensive statement, applying to everything we do. After all, taking care of the Earth is also very much God's work.

In facing environmental problems, therefore, we are called to exercise stewardship in as thorough a manner as possible, looking to God for the ability to carry it out. For any situation there are bound to be limitations to our knowledge and our ability to control; what we are invited to do is to go into the situation in partnership with God, knowing that he can take care of those things which we cannot. A clear understanding of the responsibilities and abilities we have been given coupled with trust in God's presence and trustworthiness is the mixture that makes stewardship of the Earth an exciting and challenging activity.

Given the uncertainties that surround the science of global warming, where does the path of wisdom lie? For instance, should action be taken now or should we wait until the uncertainties are less before deciding on the right action to take? The next chapter will consider directly this question of scientific uncertainty.

A prayer of Reinhold Niebuhr, an American theologian who lived earlier this century, is appropriate to conclude this section, as we face the seemingly impossible demands and challenges of global stewardship: 'O God, give us serenity to accept what cannot be changed; courage to change what should be changed; and wisdom to distinguish the one from the other.'

FOOTNOTES

1 Al Gore, *Earth in the Balance*, Houghton Mifflin Company, 1992.

2 Rachel Carson, *Silent Spring*, Houghton Mifflin Company, 1962.

3 For more information see G. Lean, D. Hinrichsen, A. Markham, *Atlas of the Environment*, Arrow Books, 1990.

4 J.E. Lovelock, *Gaia*, OUP, 1979 and *The ages of Gaia*, OUP, 1988.

5 For more details see J.E. Lovelock, *The ages of Gaia*.

6 J.E. Lovelock, *The ages of Gaia*, p 82.

7 J.E. Lovelock *The ages of Gaia*, pp 131–33.

8 J.E. Lovelock and L. Margulis, *Tellus*, **26**, 1974, pp 1–10.

9 J.E. Lovelock, 'Hands up for the Gaia hypothesis', *Nature*, **344**, 1990, pp 100–102.

10 J.E. Lovelock *Gaia: the practical science of planetary medicine*, Gaia Books, 1991.

11 J.E. Lovelock, *The ages of Gaia*, p 212.

12 Quoted by Al Gore *op. cit.* p 259

13 Quoted by Al Gore *op. cit.* p 261

14 M.H. Khalil, 'Islam and the Ethic of Conservation', *Ethics*

15 See for instance Paul Davies, *The Mind of God*, Simon & Schuster, 1992. I have also addressed this theme in J.T. Houghton, *Does God play dice?*, IVP, 1988.

16 Stephen Hawking, *A brief history of time*, Bantam Books, 1989.

17 See for instance Paul Davies 1992 *op. cit.*, also Barrow and Tipler, *The Anthropic Cosmological Principle*, OUP, 1986.

18 Barrow and Tipler *op. cit.* and J. Gribbin & M. Rees, *Cosmic coincidences*, Black Swan, 1991.

19 Paul Davies *op. cit.*

20 For modern expositions of the Christian position see R. Elsdon, *Greenhouse Theology*, Monarch, 1992 and Colin Russell, *The Earth, Humanity and God*, UCL Press, London, 1994.

21 In the first chapter of the Judaeo-Christian scriptures, we have the description of creation: 'God saw all that he had made and it was very good' (Genesis1:31), and we are told that human beings were made 'in the image of God' (1:27).

22 Al Gore, *op. cit.*

23 Al Gore, *op. cit.* p 252.

24 Al Gore *op. cit.* p 265.

25 Al Gore *op. cit.* p 254.

26 Genesis 2:15.

27 Genesis 2:19.

28 Genesis 2:12.

29 Genesis 2:9.

30 The best-known exposition of this position is L. White Jnr in, for instance, 'The historical roots of our ecological crisis', *Science*, **155**, 1987, pp 1203–1207.

31 Genesis 1:26–28.

32 A number of injunctions were given to the Jews in the Old Testament regarding care for plants and animals and care for the land, for example Leviticus 19:23–25, Leviticus 25:1–7, Deuteronomy 25:4.

33 Genesis 3:8.

34 Genesis chapter 3.

35 Romans 8:19–21.

36 A few chapters on in Genesis (9:8–17), the basis of the relationship between God and Noah is a covenant agreement in which is included a statement of God's care for the Earth. A relationship based on covenant is also the basis of the partnership between God and the Jewish nation in the Old Testament. But, after many times when that relationship was broken, the Old Testament prophets looked forward to a new covenant based not on law but on a real change of heart (Jeremiah 31:31–34). The New Testament writers (for example Hebrews 8:10–11) see this new covenant being worked out through the life and particularly through the death of Jesus. Jesus promised his followers the Holy Spirit (John 15, 16), whose influence would enable the partnership between them and God to work. Paul, in his letters, is constantly referring to the dependent relationship which forms the basis of his own partnership with God (Galatians 2:20, Philippians 4:13) and which has been the experience of millions of Christians down the centuries.

37 Genesis chapter 41.

38 Matthew 25:14–30.

39 John 15:5.

9 *Weighing the Uncertainty*

This book is intended to present clearly the current scientific position on global warming. A key part of this presentation must concern the uncertainty associated with all parts of the scientific description, especially with the prediction of future climate change—which forms an essential consideration when decisions regarding action are being taken. However, uncertainty is a relative term; utter certainty is often not demanded on everyday matters as a prerequisite for action. Here the issues are complex; we need to consider how uncertainty is weighed against the cost of possible action.

The scientific uncertainty

In earlier chapters I explained in some detail the science underlying the problem of global warming and the scientific methods which are employed for the prediction of climate change due to the increases in greenhouse gases. The basic physics of the greenhouse effect is well understood. If atmospheric carbon dioxide concentration doubles and nothing else changes apart from atmospheric temperature, then the average global temperature will increase by about 1.2°C. That figure is not disputed among scientists.

However, the situation is complicated by feedbacks and regional variations. Numerical models run on computers are the best tools available for addressing these problems. Although highly complex and at a relatively early stage of development, climate models are already capable of giving useful information of a predictive kind. Confidence in the models comes from the considerable skill with which they reproduce present climate and its variations and also from the success of the few attempts which have been made to reproduce past climates; these latter are limited as much by the lack of data as by the inadequacies of the models.

However, model limitations remain, which give rise to uncertainty (see box on p 130). The predictions presented in Chapter 6 reflected these uncertainties, the largest of which are due to the models' failure to deal adequately with clouds and with the effects of the ocean circulation. Factors—such as the regional patterns of changes in rainfall—which most influence the impact of climate change are as yet the most uncertain.

With uncertainty in the basic science of climate change and in the predictions of future climate, there are bound also to be uncertainties in our assessment of the impact of climate change. As Chapter 7 shows, however, some important general statements can be made with reasonable confidence. Under a business-as-usual scenario of increasing carbon dioxide emissions next century, the rate of climate change is likely to be large, probably greater than the Earth has seen for many millennia. Many ecosystems (including human beings) may not be able to adapt easily to such a rate of change. The most noticeable impacts are likely to be on the availability of water (especially on the frequency and severity of droughts and floods) and on the distribution (though possibly not on the overall size) of global food production. Further, although most of our predictions have been limited in range to the end of next century, it is clear that by the century beyond 2100 the magnitude of the change in climate and the impacts resulting from that change are likely to be very large indeed.

Rather less confidence is placed in estimates of the likely climate change in various broadly defined regions of the world. These estimates have been coupled with studies of the sensitivity to different climates of these regions' resources, such as water and food, and have enabled some assessment of impact to be carried out. 'Local' detail (still on a larger scale than the size of many small countries) has, however,

yet to be filled in. Predictions of the detailed impact on resources for more local regions await in their turn better scientific predictions of the likely regional climate change. The absence of more certainty about local change makes it particularly hard for politicians and decision makers to know what is the appropriate and responsible action to take.

The IPCC Assessment

Because of the scientific uncertainty, it has been necessary to make a large effort to obtain the best assessment of present knowledge and to express it as clearly as possible. For these reasons the Intergovernmental Panel on Climate Change (IPCC) was set up jointly by two United Nations' bodies, the World Meteorological Organization (WMO) and the United Nations Environmental Programme (UNEP). The IPCC's first meeting in November 1988 was very timely; it was held just as strong political interest in global climate change was beginning to develop. The Panel realized the urgency of the problem and established three working groups, one to deal with the science of climate change, one with impacts and a third one to deal with policy responses.

The task of the Science Assessment Working Group, of which I have been the chairman (since 1992, the co-chairman), has been to present in the clearest possible terms our knowledge of the science of climate change together with our best estimate of the climate change next century which is likely to occur as a result of human activities. The Working Group has produced two reports[1]: the first, a comprehensive report covering the whole field in 1990; the second, in 1992, a supplementary report, published just in time for the Earth Summit at Rio de Janeiro. Previous chapters have already referred widely to these reports. I would like to say more about how they were produced.

In preparing these reports we realized from the start that if they were to be really authoritative and taken seriously, it would be necessary to involve as many as possible of the world scientific

The reasons for scientific uncertainty

The Intergovernmental Panel on climate change, in its 1990 report[2], described the scientific uncertainty as follows.

'There are many uncertainties in our predictions particularly with regard to the timing, magnitude and regional patterns of climate change, due to our incomplete understanding of:

■ sources and sinks of greenhouse gases, which affect predictions of future concentrations,

■ clouds, which strongly influence the magnitude of climate change,

■ oceans, which influence the timing and patterns of climate change,

■ polar ice-sheets which affect predictions of sea level rise.

These processes are already partially understood, and we are confident that the uncertainties can be reduced by further research. However, the complexity of the system means that we cannot rule out surprises.'

community in their production. A small international organizing team was set up at the Hadley Centre of the United Kingdom Meteorological Office at Bracknell and through meetings, workshops and a great deal of correspondence most of those scientists in the world (both in universities and government-supported laboratories) who are deeply engaged in research into the science of climate change were involved in the preparation and writing of the reports. For the first report, 170 scientists from 25 countries contributed and a further 200 scientists were involved in its peer review. For the second report, 118 scientists from 22 countries contributed and a further 380 scientists from 63 countries and 18 UN and non-governmental organizations participated in its peer review.

The summary statements of both reports were agreed in their detailed wording at plenary meetings of the Working Group, each attended by over 100 scientists representing between 40 and 50 countries. Although there was very lively discussion at both these plenary meetings, in neither case was there any dissension from the final agreed text. Most of the discussion concerned the best and most accurate wording rather than any disagreement about the scientific information. I remember, at the meeting at Guangzhou in China which agreed the text of the 1992 report, spending over three hours on the wording of one sentence. This was not because of dispute over its scientific content. The meeting was just very concerned to present that piece of information, including the degree of scientific uncertainty we felt regarding it, as clearly and as unambiguously as possible.

During the preparation of the reports, a considerable part of the debate amongst the scientists has centred on just how much can be said about the likely climate change next century. Particularly to begin with, some felt that the uncertainties were such that scientists should refrain from making any estimates or predictions for the future. However, it soon became clear that the responsibility of scientists to convey the best possible information could not be discharged without making estimates of the most likely magnitude of the change next century coupled with clear statements of our assumptions and the level of uncertainty in the estimates. Weather forecasters have a similar, although much more short-term responsibility. Even though they may feel uncertain about tomorrow's weather, they cannot refuse to make a forecast. If they do refuse, they withhold from the public most of the useful information they possess. Despite the uncertainty in a weather forecast it provides useful guidance to a wide range of people. In a similar way the climate models, although subject to uncertainty, provide useful guidance for policy.

I have given these details of the work of the Science Assessment Group in order to demonstrate the degree of commitment of the scientific community to the understanding of global climate change and to the communication of the best scientific information to the world's politicians and policy makers. After all, the problem of global environmental change is the

largest single problem facing the world scientific community. No previous scientific assessments on this or any other subject have involved so many scientists so widely distributed both as regards their countries and their scientific disciplines. The IPCC reports can therefore be considered as authoritative statements of the contemporary views of the international scientific community.

Since the publication of the reports the debate concerning the scientific findings has continued in the world's press. Many have commented favourably on their clarity and accuracy. A few scientists have criticized because they feel the reports have insufficiently emphasized the uncertainties; others have expressed their disappointment that they have not spelt out the potential dangers to the world more forcefully. The scientific debate continues as indeed it must; argument and debate are intrinsic to the scientific process.

In describing the work of the IPCC I have concentrated on its work on basic science. This has been because of my close involvement with this part of the assessment and also because it is on the basic scientific findings that the work on impacts and response strategies must rest. Much important preparatory work was carried out by the working groups dealing with these areas in their first reports; for instance, the understanding of the impacts of climate change outlined in Chapter 7 relies heavily on the work of the Impacts Working Group. As confidence in the scientific basis of climate change becomes more established, the IPCC's work on impacts and response strategies is developing rapidly. The next comprehensive report to be published by the IPCC in 1995 will reflect a great deal of detailed work now under way in these areas.

In the presentation of the IPCC Assessments to politicians and policy makers, the degree of scientific consensus which has been achieved has been of great importance in persuading them to take seriously the problem of global warming and its impact. In the run-up to the United Nations Conference on Environment and Development (UNCED) at Rio de Janeiro in June 1992, the fact that they accepted the reality of the problem led to the formulation of the Climate Convention. It has often been commented that without the clear message which came from the world's scientists, orchestrated by the IPCC, the world's leaders would

Space observations of the climate system

For forecasting the weather round the world—for airlines, for shipping, for many other applications and for the public—meteorologists rely extensively on observations from satellites. Under international agreements, five geostationary satellites are spaced around the equator for weather observation; moving pictures from them have become familiar to us on our television screens. Information from polar orbiting satellites flown by the United States is also available to the weather services of the world to provide input into computer models of the weather and to assist in forecasting (see for instance Fig. 5.4).

These weather observations provide a basic input to climate models. But for climate prediction and research, observations from other components of the climate system, in particular from the oceans, are required. The ERS-1 satellite launched by the European Space Agency in 1991 is an example of a new generation of large satellites in which the latest techniques are directed to observing the Earth. It carries a payload particularly aimed at ocean observation (Fig. 9.1) including instruments for accurate measurement of sea surface temperature, the surface wind over the oceans (by means of a radar scatterometer) and the topography of the ocean surface (by means of a radar altimeter). This latter instrument can detect changes in the mean height of areas of the ocean surface with a precision of a few centimetres, enabling ocean currents to be located and measured. In addition ERS-1 carries a synthetic aperture radar (SAR) which provides images of the Earth's surface, including for instance its ice cover, and which is able to penetrate the cover of underlying clouds.

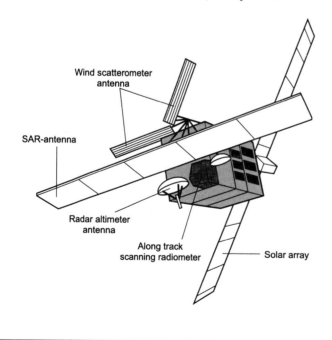

Wind scatterometer antenna

SAR-antenna

Radar altimeter antenna

Along track scanning radiometer

Solar array

FIG. 9.1 The ERS-1 satellite showing the solar array, the SAR antenna, the radar altimeter and scatterometer antennae and the along track scanning radiometer for accurate sea surface temperature measurement. The SAR antenna is 10 metres by 1 metre and the total mass of the satellite is 2.4 tonnes.

never have agreed to sign the Climate Convention.

Narrowing the uncertainty

A key question for policy makers is, 'How long will it be before the scientists are more certain about the projections of likely climate change, in particular concerning the regional and local detail?' Because of the enormous complexity of the climate system, we cannot expect the progress to be extremely rapid. Better models are needed, which in their turn will require bigger and faster computers. Above all, much better observations of all aspects of the climate system are required to describe climate variations as they occur and to calibrate and validate climate models.

In the atmosphere, clouds and all aspects of the hydrological (water) cycle need to be better observed. And it is the major oceans of the world, which cover a large fraction of the Earth's surface and which are particularly poorly monitored at the present, where observations with much higher accuracy and more complete coverage are urgently needed. To this end, new methods of observing the ocean surface from space vehicles have recently been developed and new means of observing the interior of the ocean are urgently being pursued. But not only are better physical measurements required. To be able to predict the detailed increases of greenhouse gases in the atmosphere, the problems of the carbon cycle must be unravelled; for this much more comprehensive measurements of the biosphere in the ocean as well as that on land are needed.

Stimulated by internationally organized observing programmes

such as the Global Climate Observing System (GCOS), space agencies around the world have plans in place for deploying Earth-orbiting platforms around the turn of the century which will make many new observations relevant to the problems of climate change (see box).

The vast increase seen recently in the public and political interest in the problem of climate change has stimulated a large increase in scientific activity. Through this we can expect a steady progress in our understanding. However, because the new observations mentioned above will take some years to realize and to analyze it will probably be the best part of a decade before large strides towards more complete certainty can be made and before the required detail on the regional and local scales can be provided.

Alongside the increased understanding and more accurate predictions of likely climate change coming from the community of natural scientists, much more effort is also going into studies of different human activities and how they might be affected. Much better quantification of the impacts of climate change will result from these studies. Economists and other social scientists are also beginning to carry out detailed work on possible response strategies and the economic and political measures which will be necessary to achieve them. A lot of progress can be expected during the next decade.

Sustainable development

So much for uncertainty in the science of global warming. But how does this uncertainty map on to the world of political decision making?

Sustainable development

A number of definitions of sustainable development have been produced. The following two well capture the idea.

According to the Bruntland Commission Report *Our Common Future* presented in 1987, sustainable development is 'meeting the needs of the present without compromising the ability of future generations to meet their own needs'.

A more detailed definition is contained in the White Paper *This Common Inheritance*, published by the United Kingdom Department of the Environment in 1990: 'sustainable development means living on the Earth's income rather than eroding its capital' and 'keeping the consumption of renewable natural resources within the limits of their replenishment'. It recognizes the intrinsic value of the natural world. explaining that sustainable development 'means handing down to successive generations not only man-made wealth (such as buildings, roads and railways) but also natural wealth, such as clean and adequate water supplies, good arable land, a wealth of wildlife and ample forests'.

The United Kingdom Government's first strategy report on sustainable development, issued in January 1994[3], defined four principles which should govern necessary collective action:

■ Decisions should be based on the best possible scientific information and analysis of risks.

■ Where there is uncertainty and potentially serious risks exist, precautionary action may be necessary.

■ Ecological impacts must be considered, particularly where resources are non-renewable or effects may be irreversible.

■ Cost implications should be brought home directly to the people responsible—the 'polluter pays' principle.

One of the remarkable movements of the last few years is the way in which problems of the global environment have moved up the political agenda. In her speech at the opening in 1990 of the Hadley Centre at the United Kingdom Meteorological Office, Baroness Thatcher explained our clear responsibility to the environment. 'We have a full repairing lease on the Earth. With the work of the IPCC, we can now say we have the surveyor's report; and it shows there are faults and that the repair work needs to start without delay. The problems do not lie in the future, they are here and now: and it is our children and grandchildren, who are already growing up, who will be affected.' Many other politicians have similarly expressed their feelings of responsibility for the global environment. Without this deeply felt and widely held concern, the UNCED conference at Rio, with environment as the number one item on its agenda, could never have taken place.

But, despite its importance, even when concentrating on the long term, the environment is only one of many considerations politicians must take into account. For developed countries, the maintenance of living standards, full employment (or something close to it) and economic growth have become dominant issues. Many developing countries are facing acute problems in the short term: basic survival and large debt repayment; others, under the pressure of large increases in population, are looking for rapid industrial development.

A balance has to be struck between the provision of necessary resources for development and the long term need to preserve the environment. John Major, the Prime Minister of the United Kingdom, put it like this in his speech at the Rio Conference in June 1992. 'We have to find a balance between the needs of people and the environment in which they live. We have to find a balance between the exploitation of that environment which is vital to *peoples'* survival and the conservation of that environment which is vital to *its* survival. We have to find a balance between the needs of the living and our obligations to future generations.'

That is why the Rio conference was about Environment and Development. The formula which links the two is called sustainable development (see box)—development which does not carry with it the overuse of irreplaceable resources or irreversible environmental degradation.

The idea of sustainable development echoes what was said in Chapter 8, when addressing more generally the relationship of humans to their environment and especially the need for balance and harmony. The Climate Convention signed at the Rio Conference also recognized the need for this balance. In the statement of its objective (see box on p 144), it states the need for stabilization of greenhouse gas concentrations in the atmosphere. It goes on to explain that this should be at a level and on a time-scale such that ecosystems are allowed to adapt to climate change naturally, that food production is not threatened and that economic development can proceed in a sustainable manner.

Why not wait and see?

In the light of the scientific uncertainty, it is often argued that the case is not strong enough for any action to be taken now. What we should do is to obtain as quickly as possible, through appropriate research programmes, much more precise information about future climate change and its impact. We would then, so the argument goes, be in a much better position to decide on relevant action.

It is true that more accurate information is urgently needed so that decisions can be better informed. But it does not follow that no sensible action can be taken now.

In the first place, quite a lot is already known—enough to scope the problem as a whole. There is general consensus amongst scientists about the most likely overall magnitude of climate change and there are good indications about its probable impact. Although we are not yet very confident regarding detailed predictions, enough is known to realize that the rate of climate change due to increasing greenhouse gases will almost certainly pose a large problem to the world. It will hit some countries much more than others. Those worst hit are likely to be those in the developing world that are least able to cope with it. The climate of some countries may actually improve. But in a world where there is increasing interdependence between nations, no nation will be immune from the effects.

Secondly, the time-scale of atmospheric response is long. Carbon dioxide emitted into the atmosphere today will contribute to the increased concentration of this gas and the associated climate change for over a hundred years. The more that is emitted now, the more difficult it will be to reduce atmospheric carbon dioxide concentration to the levels which will eventually be required. Further, the time-scale for human response is also long. The power stations which will produce our energy in thirty or forty years' time are being planned and built today. The demands which are likely to be placed on them because of concerns about global warming need to be brought into the planning process now.

Thirdly, some of the actions which are appropriate to combat global warming are good for other reasons as well. In particular, as was pointed out in Chapter 8, humans are far too profligate in their use of the world's resources. Fossil fuels are burnt and minerals are used, forests are cut down and soil is allowed to be eroded, without any serious thought of the needs of future generations. The imperative of the global warming problem will help us to use the world's resources in a more sustainable way.

Fourthly, as subsequent chapters will show, some of the important actions that can be taken now are not high cost actions. In fact, some of them will not only save resources but also make financial savings and improve economic performance. Other useful actions can be taken without any large economic impact.

The Precautionary Principle

Some of these arguments for action are applications of what is often called the Precautionary Principle, one of the basic principles which was included in the Rio Declaration at the Earth Summit in June 1992.

Principle 15 in the Declaration reads, 'In order to protect the environment, the precautionary approach shall be widely applied by States according to their capabilities. Where there are threats of serious or irreversible damage, lack of full scientific certainty shall not be used as a reason for postponing cost-effective measures to prevent environmental degradation.' A similar statement is contained in article 3 of the Climate Convention (see box on p 144).

We often apply the Precautionary Principle in our day-to-day living. We take out insurance policies to cover the possibility of accidents or losses; we carry out precautionary maintenance on housing or on vehicles, and we readily accept that in medicine prevention is better than cure. In all these actions we weigh up the cost of insurance or other precautions against the possible damage and conclude that the investment is worthwhile. The arguments are very similar as the Precautionary Principle is applied to the problem of global warming.

In taking out an insurance policy we often have in mind the possibility of the really unexpected. Although covering ourselves for the most unlikely happenings is not our main reason for taking out the insurance, our peace of mind is considerably increased if the policy includes these improbable events.

In a similar way, in arguing for action concerning global warming, some have strongly emphasized the need to guard against the possibility of surprises. They point out that, because of positive feedbacks which are not yet well understood[4], the increase of some greenhouse gases could be much larger than is currently predicted. They also point to the evidence that rapid changes of climate have occurred in the past (Fig. 4.6 and 4.7) possibly because of dramatic changes in ocean circulation; they could presumably occur again.

The risk posed by such possibilities is impossible to assess. It is, however, salutary to call attention to the discovery of the ozone 'hole' over Antarctica in 1985. Scientific experts in the chemistry of the ozone layer were completely taken by surprise by that discovery. In the years since its discovery, the 'hole' has substantially increased in depth. Resulting from this knowledge, international action to ban ozone-depleting chemicals has progressed much more rapidly. The lesson for us here is that the climate system may be more vulnerable to disturbance than we have often thought it to be. When it comes to future climate change, it would not be prudent to rule out the possibility of surprises.

However, insurance companies trade on our fear of the unexpected in selling their policies, and when faced with substantial uncertainty it is easy to home in on the possibility of the unknown, especially the more devastating possibilities. I would not want to advocate that the possibility of surprises should feature too prominently in the argument for action regarding global warming; other reasons for action exist where the uncertainty is less. On the other hand, even if the risks cannot easily be quantified, neither would it be responsible to ignore entirely unexpected possibilities in weighing the action that is necessary. As in the other

areas we have been discussing, the right balance must be struck.

An argument which is sometimes advanced for doing nothing now is that by the time action is really necessary, more technical options will be available. By acting now, we might foreclose their use. Any action taken now must, of course, take into account the possibility of helpful technical developments. But the argument also works the other way. The thinking and the activity generated by considering appropriate actions now and by planning for more action later will itself be likely to stimulate the sort of technical innovation which will be required.

While speaking of technical options, I should briefly mention possible options to counteract global warming by the artificial modification of the environment (sometimes referred to as geoengineering[5]). A number of proposals for 'technical fixes' of this kind have been put forward, for instance: the installation of mirrors in space to cool the Earth by reflecting sunlight away from it; the addition of dust to the upper atmosphere to provide a similar cooling effect; the fertilization of the ocean with iron to assist in the absorption of carbon dioxide; and the alteration of cloud amount and type by adding cloud condensation nuclei to the atmosphere. None of these have been demonstrated either to be feasible or effective. Further, they suffer from the very serious problem that none of them would exactly counterbalance the effect of increasing greenhouse gases. As has been shown, the climate system is far from simple. The results of any attempt at large-scale climate modification could not be perfectly predicted and might not be what is desired. With the present state of knowledge, artificial climate modification along any of these lines is not an option that need be considered.

The conclusion from this section is that to 'wait and see' would be an inadequate and irresponsible response to what we know. The Climate Convention signed in Rio (see box on p 144) recognized that action needs to be taken now. Just what that action should be will be the subject of the next chapter.

Some global economics

So far in this chapter, our attempt to balance uncertainty against the need for action has been considered in terms of issues. Is it possible to carry out the weighing in terms of cost? In a world which tends to be dominated by economic arguments, quantification of the costs of action against the likely costs of the consequences of inaction must at least be attempted. It is also helpful to put these costs in context by comparing them with other items of global expenditure.

At the end of Chapter 7 estimates of the cost to the world of global warming, both in terms of the damage which it is likely to cause and the cost of adapting to it, were presented. Although, at the present stage of knowledge, these estimates are bound to be crude, nevertheless they give a feel for the likely range of cost.

For the IPCC business-as-usual scenario, the cost estimates for the period up to about 2050 expressed as an annual sum are of the order of 1 or 2 per cent of the Gross World Product (GWP). The cost to developing countries, because of

their greater vulnerability to climate change and because a greater proportion of their expenditure is dependent on activities such as agriculture and water, is likely to be at least twice that to developed countries. Since the main contribution to global warming will come from carbon dioxide, it is useful to express these estimates of cost in terms of the cost per tonne of carbon as carbon dioxide emitted from human activities, which totals about 7,500 million tonnes per annum. If the estimated cost of damage is 1 per cent of GWP (about 200,000 million US$), the cost per tonne of carbon is about $25; for 2 per cent of GWP, it is about $50; for 4 per cent of GWP, it is about $100.

The alternative to adapting to global warming and bearing the cost is to avoid or mitigate its effects by reducing greenhouse gas emissions, in particular the emissions of carbon dioxide. A number of studies of the cost of avoidance and mitigation have been carried out, which again have been analyzed by William Cline[5]. The reduction of emissions would mainly have to be achieved in the energy sector. In the absence of innovative technical change the cost of this would be quite high, perhaps as much on average as 3 per cent of GWP to maintain emissions at their 1990 level and considerably more than that to reduce emissions below the 1990 level[7]. However, as the next chapter will explain, there are already clear technical possibilities for energy savings and conservation and for new sources of supply which do not emit greenhouse gases. When these are taken into account, it is found that an early reduction of emissions from their business-as-usual level, perhaps by as much as 25 per cent, is achievable at little or no cost.

Further, a review of estimates of the cost of achieving the more drastic reductions required to stabilize carbon dioxide concentrations before the end of the century and hence eventually to eliminate further global warming comes up with figures of 2 per cent of GWP or less—consistent with a recent analysis commissioned by the United Nations Environment Programme which concludes that the required 'long-term reduction in projected global fossil fuel carbon dioxide emissions ... could have an impact on long-term GWP ranging from insignificant to a maximum 2.5 per cent reduction'[8].

One of the most recent studies with economic models which allow for technical advance has examined the difference in the likely cost of various pathways of achieving stabilization of carbon dioxide concentration by the year 2100[9]. Stabilization at 400ppmv, a little above today's concentration, would required drastic changes now and would be expensive—an annual cost of about 3 per cent of GWP is estimated. But the optimum path for stabilization at 450ppmv would cost much less—annually about 1 per cent of GWP—and stabilization at 500ppmv would cost less still— about 0.75 per cent of GWP annually is estimated. For the latter the emission scenario is close to that of the World Energy Council scenario C which was presented in Fig. 3.5.

Such studies provide a useful guide, especially when comparing different courses of action, although it must be emphasized that estimates for so far ahead are bound to possess a lot of uncertainty. If anything, they

are likely to be on the high side. Inevitably, it is extremely difficult to peer into the crystal ball of technical development and almost any attempt to do so is likely to underestimate its future potential.

Such 'weighing' of the economics as has been possible so far therefore brings three messages. The first is that early action at small cost can be undertaken now to reduce emissions and to slow the rate of change. The second is that, in the longer term, as presently estimated, the cost in economic terms of mitigation and avoidance is of similar order of magnitude in global average terms as the likely cost of the damage from global warming or of adapting to it. The third message is that drastic action now to stabilize carbon dioxide concentration at close to today's level within the next few decades—and therefore largely to eliminate climate change now—is likely to be expensive compared to the cost of action to stabilize the carbon dioxide concentration on a somewhat longer time-scale (say by the end of next century) and at a somewhat higher level. Since the weighing of economic factors is an important consideration for politicians and policy makers, work to refine these estimates needs to be pursued with urgency.

In considering both the cost of the impact of global warming and the cost of adaptation or of mitigation a figure of about 1 or 2 per cent of GWP has been reached. It is interesting to compare this with other items of expenditure in national or personal budgets. In the United Kingdom about 5 per cent of income is spent on the supply of primary energy (basic fuel such as coal, oil and gas and fuel for electricity supply), about 6 per cent on health and 4 to 5 per cent on defence. It is, of course, very clear that global warming is strongly linked to energy production—it is largely because of the way energy is provided that the problem exists—and this subject will be expanded in the next chapter. But the impacts of global warming also have implications for health—such as the possible spread of disease—and for national security—for example, the possibility of wars fought over water, or the impact of large numbers of environmental refugees. Any thorough consideration of the economics of global warming needs therefore to assess the strength of these implications and to take them into account in the overall economic balance.

So far, on the global warming balance sheet we have estimates of costs and of benefits or drawbacks. What we do not have as yet is a capital account. Valuing human-made capital is commonplace, but in the overall accounting we are attempting, 'natural' capital must clearly be valued too. By 'natural' capital I mean, for instance, natural resources which may be renewable (such as a forest) or non-renewable (such as coal, oil or minerals)[9]. Their value is clearly more than the cost of exploitation or extraction.

Other items, some of which were mentioned at the end of Chapter 7, such as natural amenity and the value of species, can also be considered as 'natural' capital. I have argued (Chapter 8) that there is intrinsic value in the natural world— indeed, the value and importance of such 'natural' capital is increasingly recognized. The difficulty is that it is neither possible nor appropriate to express much of this value in money.

In summary, therefore, the items in the overall global warming balance sheet are:

■ estimates (with considerable uncertainty) of cost (for those items which can be quantified) of the likely impacts of global warming under a business-as-usual scenario, which are typically of the order of 1 to 2 per cent of GWP annually and are at least twice as great in these percentage terms for developing countries as for developed countries

■ estimates of the cost of mitigation and avoidance of global warming— again, typically of the order of 1 per cent of GWP, although the likely cost of drastic and immediate action to prevent any further climate change is likely to be larger

■ impacts of global warming which affect human amenity and 'natural' capital or which have implications for national security, all of which are difficult if not impossible to value in money terms.

Some adaptation to climate change is, of course, inevitable because of the climate change to which we are already committed. There is already international acceptance that some action to mitigate global warming is necessary. The next chapter will consider how this mitigation is viewed in the wider context of the requirement for sustainable development.

FOOTNOTES

1 *Climate Change, The IPCC Scientific Assessment*,
eds J.T. Houghton, G.J. Jenkins and J.J. Ephraums, CUP,
1990, and *Climate Change 1992, The Supplementary
Report to the IPCC Scientific Assessment*,
eds J.T. Houghton, B.A. Callendar, S.K. Varney, CUP, 1992.

2 *Climate Change, The IPCC Scientific Assessment*,
eds J.T. Houghton, G.J. Jenkins and J.J. Ephraums, CUP,
1990, p 365; Executive Summary p xii.

3 *Sustainable Development: the UK Strategy*, London,
HMSO, Cm 2426, 1994, p 7.

4 See Chapter 4

5 Reviewed in *Policy implications of greenhouse warming*,
National Academy Press, Washington DC, 1992, pp 433–64.

6 W.R. Cline, *The economics of global warming*, Institute
for International Economics, Washington DC, 1992, p 399.

7 W.R.Cline, *op. cit.* p 191.

8 *UNEP Greenhouse Gas Abatement Costing Studies:
Phase 1 Report*, United Nations Environment Programme,
1992.

9 R. Richels and J. Edmonds, 'The economics of
stabilizing atmospheric CO_2 concentrations', submitted to
Nature, 1993.

10 For a discussion of this issue see H.E. Daly, 'From
empty-world economics to full-world economics: a
historical turning point in economic development' in *World
Forests for the Future*, eds K. Ramakrishna and
G.M.Woodwell, Yale University Press, 1993, pp 79–91.

10 *Action to Slow and Stabilize Climate Change*

Following *the awareness of the problems of climate change aroused by the IPCC scientific assessment, the necessity of international action has been recognized. In this chapter I address the forms that action could take.*

The Climate Convention

The United Nations Framework Convention on climate change signed by over 160 countries at the United Nations Conference on Environment and Development held in Rio de Janeiro in June 1992 has set the agenda for action to slow and stabilize climate change. The signatories to the Convention (some of the detailed wording is presented in the box below) recognized the reality of global warming, recognized also the uncertainties associated with current predictions of climate change, agreed that action to mitigate the effects of climate change needs to be taken and pointed out that developed countries should take the lead in this action.

The Convention mentions one particular aim concerned with the relatively short term and one far reaching objective. The particular aim is that countries should take action to return greenhouse gas emissions, in particular those of carbon dioxide, to their 1990 levels by the year 2000. The long term objective of the Convention,

expressed in Article 2, is that the concentrations of greenhouse gases in the atmosphere should be stabilized 'at a level which would prevent dangerous anthropogenic interference with the climate system', the stabilization to be achieved within a time-frame sufficient to allow ecosystems to adapt naturally to climate change, to ensure that food production is not threatened and to enable economic development to proceed in a sustainable manner. In setting this objective, the Convention has recognized that it is only by stabilizing the concentration of greenhouse gases (especially carbon dioxide) in the atmosphere that we can halt the rapid climate change which is expected to occur with global warming.

The following paragraphs will first consider the immediate actions which are possible to begin to meet the requirements of the Convention. Some of these are under way and in some cases agreements already exist. Further actions necessary to satisfy the Convention's objective to stabilize greenhouse gas concentrations will then be considered.

Stabilization of emissions

The target for short-term action proposed by the Climate Convention is that, by the year 2000, greenhouse gas emissions should be brought back to no more than 1990 levels. In the run-up to the Rio conference, before the Climate Convention was formulated, many developed countries already announced their intention to meet such a target at least for carbon dioxide. They would do this mainly through energy-saving measures and through switching to fuels such as natural gas which, for the same energy production, generates 40 per cent less carbon dioxide than coal and 30 per cent less than oil. More detail of these energy-saving measures are given in the next chapter, which is devoted to a discussion of future energy needs and production.

The question is bound to be asked: what happens after the year 2000? For the countries which have entered into commitments for the year 2000, will emissions continue to be stabilized or will they be allowed to grow again? Some countries have looked a little further into the future; for instance, Germany and Denmark have made commitments to reduce emissions by 20 or 25 per cent of 1990 levels by the year 2005—although they have not spelt out in detail how this is to be achieved—but most countries have said virtually nothing about the longer term. Strategies for the longer term, to begin to meet the requirements of the Climate Convention Objective, will be discussed later on in this chapter.

The Montreal Protocol

The chlorofluorocarbons (CFCs) are greenhouse gases whose emissions into the atmosphere are already controlled under the Montreal Protocol on ozone depleting substances. This control has not arisen because of their potential as greenhouse gases, but because they deplete atmospheric ozone (see Chapter 3). Although emissions of CFCs have fallen sharply during the last few years, their concentration continues to increase, because of their long life in the atmosphere. However, the phase-out of their manufacture in industrialized countries by 1996 and in developing countries by 2006 as required by the 1992 amendments to the Montreal Protocol, will ensure that the profile of their atmospheric concentration will turn around and begin to decline before the end of the century. However, because their decline in the atmosphere will be slow, it will be a century or more before their contribution to global warming is reduced to a negligible amount.

The replacements for CFCs, the HCFCs, which are also greenhouse gases—though less potent than the CFCs—are required to be phased out by 2030. It will probably be close to that date before their atmospheric concentration stops rising and begins to decline. Any substantial growth in HFCs, which are greenhouse gases but not ozone-depleting (see Chapter 3) needs to be watched. But international agreements now exist for the CFCs and most of the related species which contribute to the greenhouse effect, which will bring about the stabilization required by the Climate Convention.

Forests

We now turn to the situation of the world's forests and the contribution that they can make to the mitigation of global warming; action here can

Some extracts from the United Nations Framework Convention on climate change, signed by over 160 countries in Rio de Janeiro in June 1992

First, some of the paragraphs in its preamble, where the parties to the Convention:

CONCERNED that human activities have been substantially increasing the atmospheric concentration of greenhouse gases, that these increases enhance the natural greenhouse effect, and that this will result on average in an additional warming of the Earth's surface and atmosphere and may adversely affect natural ecosystems and humankind,

NOTING that the largest share of historical and current global emissions of greenhouse gases has originated in developed countries, that per capita emissions in developing countries are still relatively low and that the share of global emissions originating in developing countries will grow to meet their social and development needs,

RECOGNIZING that various actions to address climate change can be justified economically in their own right and can also help in solving other environmental problems,

RECOGNIZING that low-lying and other small island countries, countries with low-lying coastal, arid and semi-arid areas or areas liable to floods, drought and desertification, and developing countries with fragile mountainous ecosystems are particularly vulnerable to the adverse effects of climate change,

AFFIRMING that responses to climate change should be coordinated with social and economic development in an integrated manner with a view to avoiding adverse impacts on the latter, taking into full account the legitimate priority needs of developing countries for the achievement of sustained economic growth and the eradication of poverty,

DETERMINED to protect the climate system for present and future generations,

have AGREED as follows:

The Objective of the Convention is contained in Article 2 and reads as follows:

'The ultimate objective of this Convention and any related legal instruments that the Conference of the Parties may adopt is to achieve, in accordance with the relevant provisions of the Convention, stabilization of greenhouse gas concentrations in the atmosphere at a level that would prevent dangerous anthropogenic interference with the climate system. Such a level should be achieved within a time frame sufficient to allow ecosystems to adapt naturally to climate change, to ensure that food production is not threatened and to enable economic development to proceed in a sustainable manner.'

Article 3 deals with principles and includes agreement that the Parties 'take precautionary measures to anticipate, prevent or minimize the causes of climate change and mitigate its adverse effects. Where there are threats of serious or irreversible damage, lack of full scientific certainty should not be used as a reason for postponing such measures, taking into account that policies and measures to deal with climate change should be cost-effective so as to ensure global benefits at the lowest possible cost.'

Article 4 is concerned with Commitments.

In this article, each of the signatories to the Convention agreed 'to adopt national policies and take corresponding measures on the mitigation of climate change, by limiting its anthropogenic emissions of greenhouse gases and protecting and enhancing its greenhouse sinks and reservoirs. These policies and measures will demonstrate that developed countries are taking the lead in modifying longer-term trends in anthropogenic emissions consistent with the objective of the Convention, recognizing that the return by the end of the present decade to earlier levels of anthropogenic emissions of carbon dioxide and other greenhouse gases not controlled by the Montreal Protocol would contribute to such modification...'

Each signatory also agreed 'in order to promote progress to this end...to communicate...detailed information on its policies and measures referred to above, as well as on its resulting projected anthropogenic emissions by sources and removals by sinks of greenhouse gases not covered by the Montreal Protocol ... with the aim of returning individually or jointly to their 1990 levels these...emissions...'

easily be taken now and is commendable for many other reasons.

Many of the largest and most critical forested areas are in the tropics. In recognition of the extremely valuable resource these areas represent, the developing countries where they lie are beginning to concentrate seriously on the management of their forests, on limiting the extent of deforestation or planning for substantial afforestation. Other large areas of forest lie at higher latitudes where developed countries can also take useful action to contribute to the alleviation of the problem of global warming.

During the last two decades, the additional needs of the increasing populations of developing countries for agricultural land and for fuelwood together with the rise in demand for tropical hardwoods by developed countries has led to a worrying rate of loss of forest in tropical regions. In many tropical countries the development of forest areas has been the only hope of subsistence for many people. Unfortunately, because the soils and other conditions were often inappropriate, much of this forest

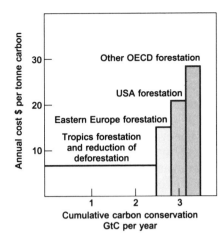

FIG. 10.1 **Average annual costs by region for a programme of afforestation to fix carbon[1].**

clearance has not led to sustainable agriculture but to serious land and soil degradation[2].

Measurements on the ground and observations from orbiting satellites have been combined to provide estimates of the area of tropical forest lost. Over the decade of the 1980s the average loss has been about 1 per cent per year (Table 10.1); in some areas the loss has been considerably higher. Such rates of loss cannot be sustained if much forest is to be left in fifty or a hundred years' time. The loss of forests is damaging, not only because of the ensuing land degradation but also because of the contribution that loss makes to global warming. There is also the dramatic loss in

TABLE 10.1 **Estimates, from the United Nations Food and Agriculture Organization (FAO), of forest cover and deforestation for 87 countries in tropical regions[3]. The countries include almost all of the moist forest zone together with some dry areas. The figures are indicative and should not be taken as regional averages.**

Forest cover and deforestation for tropical regions

Continent	Forest Area 1980	Forest Area 1990	Annual deforestation 1981–90	Rate of change 1981–90 (%/year)
	(thousands of square kilometres)			
Africa	6,500	6,000	50	-0.8
Latin America and Caribbean	9,230	8,400	83	-0.9
Asia	3,210	2,750	37	-1.2
Total	18,940	17,150	170	-0.9

biodiversity (it is estimated that over half the world's species live in tropical forests) and the potential damage to regional climates (loss of forests can lead to a significant regional reduction in rainfall—see box on p 101).

For every square kilometre of a typical tropical forest there are between about 20,000 and 50,000 tonnes of biomass (total living material), containing 10,000–25,000 tonnes of carbon[4]. It is estimated that burning or other destruction from deforestation turns about two-thirds of this carbon into carbon dioxide. On this basis, from a destruction of 170,000 square kilometres, nearly 2 Gt (1 Gt = 1 gigatonne = 1,000 million tonnes) of carbon would enter the atmosphere as carbon dioxide. Although these estimates are rather crude and there are substantial uncertainties in the numbers, they tally with the IPCC estimate, quoted in Chapter 3 (see Table 3.1), of the carbon as carbon dioxide entering the atmosphere each year from land-use change (mostly deforestation) of 1.6 ± 1.0 Gt per year—a significant fraction of the total emissions of carbon dioxide into the atmosphere from human activities. Reducing deforestation can therefore make a substantial contribution to slowing the increase of greenhouse gases in the atmosphere, as well as the provision of other benefits such as guarding biodiversity and avoiding soil degradation.

So much for slowing deforestation. How about the possibilities for afforestation? For every square kilometre, a fully growing forested area fixes between 500 and 1000 tons of carbon per year[5]. To illustrate the effect of afforestation on atmospheric carbon dioxide, suppose that an area of 100,000 square kilometres, a little more than the area of the island of Ireland, were planted each year for 40 years—starting now. By the year 2035, 4 million square kilometres would have been planted, that is roughly half the area of Australia. By the time the new forests matured— between 40 and 100 years after planting (the actual period depending on the type of forest)—between about 50 and 100 Gt of carbon from the atmosphere would have been sequestered. This accumulation of carbon in the forests is equivalent to at least 10 per cent of the emissions due to fossil fuel burning in the business-as-usual scenario during the first half of the next century. It would provide a useful contribution to any required reduction in atmospheric carbon dioxide concentrations.

But is such a tree planting programme feasible and is land on the scale required available? The answer is almost certainly, yes. A study by R.A. Houghton for the tropical regions has identified land which has supported forests in the past and which is not presently being used for croplands or settlements, of an area totalling about 5 million square kilometres— in Latin America (1 million square kilometres), Asia (1 million square kilometres) and Africa (3 million square kilometres)[6]. The cost of afforestation is typically $(US)40,000 per square kilometre; it can vary from around $20,000 to $100,000 depending on the cost of the land and where it is done[7]. The annual programme we have mentioned would cost therefore about $4,000 million per year. The cost, expressed in terms of the amount of carbon fixed, averages about $10 per tonne (see Fig. 10.1). Such a figure makes the programme a potentially attractive one[8] for alleviating the rate of change of climate due to increasing greenhouse gases in the relatively short term. And only tropical regions have been considered so far. Potential for afforestation, not so great but still substantial, also exists at mid and high latitudes (Fig. 10.1).

A possible afforestation programme has been presented in order to illustrate the potential for carbon sequestration. Once the trees are fully grown, of course, the sequestration ceases. What happens then depends on the use which may be made of them. They may be for 'protection' forests, for instance for the control of erosion or for the maintenance of biodiversity; or they may be production forests, used for biofuels (see Chapter 11) or for industrial timber. If they are used for fuel, they add to the atmospheric carbon dioxide but, unlike fossil fuels, they are a renewable resource. As with the rest of the biosphere where natural recycling takes place on a wide variety of time-scales, carbon from wood fuel can be continuously recycled through the biosphere and the atmosphere.

Reduction in the sources of methane

Methane is a less important greenhouse gas than carbon dioxide, contributing perhaps 15 per cent to the present level of global warming. The stabilizing of its atmospheric concentration would contribute a small but significant amount to the overall problem. Because of its much

shorter lifetime in the atmosphere (about 10 years compared with 100–200 years for carbon dioxide), only a relatively small reduction in the emissions of this gas, less than 10 per cent in fact, would be required to stabilize its concentration at the current level.

Referring to Table 3.2 which lists the various sources of methane, there are three sources arising from human activities which could rather easily be reduced at small cost[9]. First, methane emission from biomass burning would perhaps be halved if deforestation were drastically curtailed.

Secondly, methane production from landfill sites could be cut by at least a half if more waste were recycled, if arrangements were made for the collection of methane gas (it could then be used for energy production or if the quantity were insufficient it could be flared) and if more waste were used for energy generation by incineration. Waste management policies in many countries already include the encouragement of such measures.

Thirdly, the leakage from natural gas pipelines from mining and other parts of the petrochemical industry could at little cost (possibly even at a saving in cost) be reduced by, say, one third. An illustration of the scale of the leakage is provided by the suggestion that the closing down of some Siberian pipelines because of the major recession in Russia has been the cause of the fall in the growth of methane concentration in the atmosphere from 1992 to 1993. Improved management of such installations could markedly reduce leakage to the atmosphere, perhaps by as much as one quarter overall.

Reduction of these three sources as suggested could result in a reduction of methane emissions by some 60 million tonnes per annum which would be adequate, even allowing for some small increases in emissions from other sources, to stabilize the concentration of methane in the atmosphere at about the current level. Put another way, the reduction in methane emissions from these sources would be equivalent to a reduction in annual carbon dioxide emissions producing about one third of a gigatonne of carbon[10] or a little less than 5 per cent of total greenhouse gas emissions—a useful contribution towards the solution of the global warming problem.

Stabilization of carbon dioxide concentrations

We now turn to consider the stabilization of the atmospheric concentration of carbon dioxide, the most important of the greenhouse gases that result from human activities. Under a business-as-usual scenario, emissions will continue to increase throughout next century with no sign of abatement. The concentration of carbon dioxide also rises continuously through that period; it increases to about two and a half times its pre-industrial value by the year 2100. Under this sort of scenario of continually increasing emissions no stabilization of carbon dioxide concentration, and hence no stabilization of climate, is in sight.

What sort of emissions scenario would stabilize the carbon dioxide concentration? The commitment of some nations to ensure that carbon dioxide emissions in the year 2000 are no greater than they were in 1990 has already been mentioned. Suppose, after the year 2000, all

countries kept their carbon dioxide emissions constant, would that be enough?

Stabilizing concentrations is, however, very different from stabilizing emissions. With constant emissions after the year 2000, the concentration in the atmosphere would continue to rise and would approach 500ppmv by the year 2100 (Fig. 3.6). After that the carbon cycle models predict that the carbon dioxide concentration would still continue to increase, although more slowly than at present, but show no sign of stabilization for at least the next four hundred years[11]— although the uncertainty in using the models to project that far ahead is large. In the terms of the objective of the Climate Convention, both the level of climate change resulting from such a scenario and the time taken to achieve climate stability are almost certainly unacceptable.

If we are to reach stabilization of the carbon dioxide concentration in the atmosphere by the end of next century, emissions will eventually have to be reduced below current levels. As Chapter 3 showed (Fig. 3.6), the World Energy Council ecologically driven scenario (Scenario C) provides an example of a pathway to stabilization on that time-scale. Under that scenario, global carbon dioxide emissions grow by about 10 per cent by the year 2050; they then fall by a factor of two by the end of next century. Up to the year 2020 emissions in the developing world are allowed to approximately double, while those for developed countries fall by about 2 per cent per annum.

As the World Energy Council point out in their report, achievement of such a scenario will be far from easy. It requires three essential ingredients. The first is an aggressive emphasis on energy saving and conservation. Much here can be achieved at zero net cost or even at a cost saving. Though much energy conservation can be shown to be economically advantageous, it is unlikely to be undertaken without significant incentives. However, it is clearly good in its own right, it can be started in earnest now and it can make a significant contribution to the reduction of emissions and the slowing of global warming. The second ingredient is an emphasis on the development of appropriate renewable energy sources and the third is the transfer of technologies to developing countries which will enable them to apply the most appropriate and the most efficient technologies to their industrial development, especially in the energy sector. The next chapter will give more detail about these energy matters.

The World Energy Council scenario C is one example of a future energy scenario which takes account of the reality of global warming. It is, of course, only one of many such scenarios which could be considered. How are we going to decide the best path to follow so far as energy strategy, for instance, is concerned?

I turn for guidance to the Climate Convention which recognizes in its basic objective (see box on p 144) the need to stabilize the concentrations of the important greenhouse gases. It goes on to say that the level of stabilization and the time-scale for its achievement should be such that ecosystems should be able to adapt naturally, that food production must not be threatened and that economic development can proceed in a sustainable manner. We do not yet know enough to pick precisely the level or the time-scale under the criteria the Climate Convention is prescribing, but perhaps already some limits can already be set. On the one hand, the long life of carbon dioxide in the atmosphere provides quite a severe constraint on the possibilities for the growth of emissions. For instance, any scenario with substantial growth in global emissions over the next half century is unlikely to lead to an acceptable level (using the criteria for ecosystem adaptation and food production) or time-scale for the stabilization required. On the other hand, the drastic reduction in emissions, by about a half, which would have to take place almost immediately to stabilize at today's concentration of carbon dioxide would also be unacceptable; it could not be achieved without severe economic disruption.

Summary of the action required

This chapter has suggested some actions that can be taken to slow climate change and ultimately to stabilize it as is required by the internationally agreed Climate Convention.

Some actions have already been taken which have an effect on global emissions of greenhouse gases, namely:

■ the reduction by some countries of carbon dioxide emissions in the year 2000 to 1990 levels and

■ the provisions of the Montreal Protocol regarding the emissions of CFCs and CFC substitutes.

149

Other actions which can be taken now to slow climate change, which can be done at little or no net cost and which are good to do for other reasons are the following:

■ a reduction of deforestation,

■ a substantial increase in afforestation,

■ some easy-to-do reductions in methane emissions and

■ an aggressive increase in energy saving and conservation measures.

For the longer term, as well as increased emphasis on these measures, the world needs to begin to follow an energy scenario which will lead to the stabilization of carbon dioxide concentration in the atmosphere, for instance one similar to the World Energy Council scenario C. How this might be achieved will be addressed in detail in the next chapter.

At this present stage of knowledge, that WEC scenario C is just an example of what may be required. It may or may not be sufficient. However, it may be noted that, to begin with, it demands, roughly speaking, constant global emissions of carbon dioxide. Really drastic reductions in global emissions do not begin until about

the middle of next century. Well before that time much more precise information about climate change should be available.

FOOTNOTES

1 Adapted from *Climate Change: The IPCC Response Strategies*, IPCC, 1990, p 101.

2 More detail in *The world environment 1972–1992*, eds M.K. Tolba, O.A. El-Kholy, Chapman and Hall, 1992, pp 157–82.

3 Quoted in *The world environment 1972–1992*, eds M.K. Tolba, O.A. El-Kholy, Chapman and Hall, 1992, p 169.

4 These figures do not include carbon in the soils. See Salati *et al.*, *Climate Change: Science, Impacts and Policy; Proceedings of the Second World Climate Conference*, eds J. Jager & H.L. Ferguson, CUP 1991, pp 391–95; also J. Leggett *et al.*, *Emission scenarios for the IPCC: an update in climate change*, IPCC, 1992, p89; a more detailed source is R.A. Houghton, *Climate Change* **19**, 1991, pp 99–118.

5 See *Climate Change: The IPCC Response Strategies*, IPCC, 1990, p 87.

6 R.A. Houghton, 'Role of forests in global warming' in *World Forests for the Future: their use and conservation*, eds K.Ramakrishna and G.M.Woodwell, Yale University Press, 1993, pp 21–58, also quoted in *Climate Change: The IPCC Response Strategies*, IPCC, 1990, p 101.

7 See *Climate Change: The IPCC Response Strategies*, IPCC, 1990, p 102.

8 Compare with the figures of 25$, 50$ or 100$ per tonne mentioned for the cost of global warming at the end of Chapter 9.

9 Further details in *Technological Options for reducing Methane Emissions, Background Document of the Response Strategies Working Group*, IPCC, January 1992.

10 This figure is calculated by multiplying the 60 million tonnes by the global warming potential for methane which, for a time horizon of 100 years, is about 7.5 (molecule CH_4 per molecule CO_2), then by $^{12}/_{16}$ to put it into tonnes of C.

11 Information from Professor Tom Wigley.

11 *Energy and Transport for the Future*

We *flick a switch and energy flows. Energy is provided so easily for the developed world that thought is rarely given to where it comes from, whether it will ever run out or whether it is harming the environment. Energy is also cheap enough that little serious attention is given to conserving it. However, if the emissions of greenhouse gases into the atmosphere are to be reduced, a large proportion of the reduction will have to occur in the energy sector. There is a need, therefore, to concentrate the minds of policy makers and indeed of everyone on our energy requirements and usage. This chapter looks at how future energy might be provided in a sustainable manner.*

World energy demand and supply

Most of the energy we use can be traced back to the sun. In the case of fossil fuels it has been stored away over millions of years in the past. If wood (or other biomass including animal and vegetable oils), hydro-power, wind or solar energy itself is

FIG. 11.1 Growth in the rate of energy use and in the sources of energy since 1860 in thousand millions of tonnes of oil equivalent (Gtoe) per year[1]. The energy shown for electricity is that used in its generation (for which the basic unit is the watt (thermal)); the efficiency of conversion into usable electrical energy (for which the basic unit is the watt (electrical)) is typically around one third.

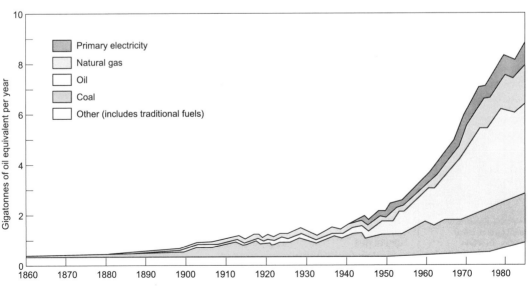

used, the energy has either been converted from sunlight almost immediately or has been stored for at most a few years. These latter sources of energy are renewable; they will be considered in more detail later in the chapter. The only common form of energy that does not originate with the sun is nuclear energy; this comes from radioactive elements which were present in the Earth when it was formed.

Until the Industrial Revolution, energy for human society was provided from 'traditional' sources— wood and other biomass and animal power. Since 1860, as industry has developed, the rate of energy use has multiplied by about a factor of 30 (Fig. 11.1), at first mostly through the use of coal followed, since about 1950, by rapidly increasing use of oil. In 1990 the world consumption of energy was about 8,730 million tonnes of oil equivalent (toe). This can be converted into physical energy units to give an average rate of energy use of about 12 million million watts (or 12 Terawatts $= 12 \times 10^{12}$ W)[2].

Great disparities exist in the amount of energy used per person in various parts of the world (Fig. 11.2). In 1990 each person in the world used on average 1.65 tonnes of oil equivalent (toe), an average consumption of energy of about 2.2 kilowatts (kW). The highest rates of energy consumption are in North America where the average citizen in 1990 used nearly 8 toe, equivalent to an average rate of consumption of about 11 kW. By contrast the average Indian used only one twentieth of that amount or about 0.4 toe, equivalent to an average rate of consumption of 0.5kW, mainly in the

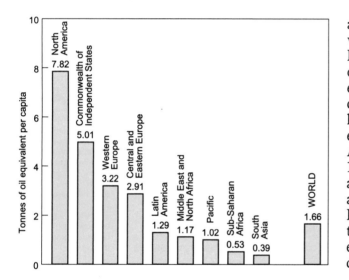

FIG. 11.2 Per capita energy use in tonnes of oil equivalent in 1990 in different regions of the world[3].

TABLE 11.1 Reserves of fossil fuels and their relation to current energy use. Data from World Energy Council[4].

Estimate of fossil fuel resources

	Proven reserves in 1990 (Gtoe)	Ratio: Reserves to current annual use (years)	Ultimately recoverable resources (Gtoe)
Coal (excluding lignite)	496	197	3400
Lignite	110	293	
Conventional oil	137	40	200
Unconventional oils (heavy crude, natural bitumen, oil shale)			600
Natural gas	108	56	220

form of traditional fuels. In fact about half the world's population rely wholly on traditional fuels and do not currently have access to commercial energy in any of its forms.

It is interesting to see how the energy we consume is used. Taking the world average, about 20 per cent of primary energy is used in transportation, about 40 per cent by industry and the remaining 40 per cent in commercial activity and in homes. It is also perhaps interesting to know how much energy is used in the form of electricity. About one third of primary energy goes to make electricity at an average efficiency of conversion of about one third. Of this electrical power about half, on average, is utilised by industry and the other half in commercial activities and in homes.

How much is spent on energy? Taking the world as a whole, the amount spent per year by the average person for the 1.65 toe of energy used, is about 5 per cent of annual income. Despite the very large disparity in incomes, the proportion spent on primary energy is much the same in developed countries and developing ones.

How are we likely to be placed over energy for the future? If we continue to generate most of our energy from coal, oil and gas, do we have enough to keep us going? Current knowledge of proven recoverable reserves (Table 11.1) indicates that known reserves of fossil fuel will meet demand for the period up to 2020 and substantially beyond. Around mid-century, if demand continues to expand, oil and gas supplies will come under increasing pressure. Further exploration will be stimulated which will lead to the exploitation of more

sources, although increased difficulty of extraction can be expected to lead to a rise in price. So far as coal in concerned, there are plentiful supplies for well over a hundred years.

Estimates have also been made of the ultimately recoverable fossil fuel reserves, defined as those potentially recoverable assuming high but not prohibitive prices and no significant bans on exploitation. Although these are bound to be somewhat speculative, they show (Table 11.1) that, at current rates of use, reserves of oil and gas are likely to be available for 100 years and of coal for more than 1000 years.

Likely reserves of uranium for nuclear power stations should also be included in this list. When converted to the same units they are believed to be at least 3,000 and possibly as high as 9,000 Gtoe, substantially greater than likely fossil fuel reserves.

Future energy projections

Many energy experts, individuals and organizations have attempted to look into the crystal ball of the future and estimate future energy demand next century and how it might be met. Any such attempts are bound to be fraught with a lot of uncertainty and to include assumptions of dubious accuracy; no one really knows how human beings will behave or what political or technical developments may occur. Nevertheless a range of possibilities must be considered in order to predict what climate change might occur. The predictions of climate change drawn up in Chapter 6 based on a business-as-usual scenario prepared under the auspices of the IPCC (scenario IS 92a) estimated future energy demand on the assumption that no controls or constraints for environmental reasons were applied. It is interesting to compare the energy assumptions of that scenario with four energy scenarios constructed by the World Energy Council (WEC)[5]. These scenarios (more details in the box) take into account likely population growth and energy sources and a realistic view of the rate of technical change. One of the scenarios, the 'high growth' case (A), assumes a higher rate of economic growth in developing countries. An 'ecologically driven' case (C) assumes that environmental pressures have a

World Energy Council and IPCC scenarios

The World Energy Council (WEC) is an international non-governmental organization with representation from all parts of the energy industry and from over 90 countries. The Council has developed four energy scenarios for the period to 2020, each representing different assumptions in terms of economic development, energy efficiencies, technology transfer and the financing of development round the world[6]. The WEC emphasizes that they have been developed to illustrate future possibilities and they should not be considered as predictions. For all four cases, following the United Nations base case, world population is assumed to grow from 5.3 thousand million in 1990 to 8.1 thousand million in 2020 (and further to 10 thousand million in 2050 and 12 thousand million in 2100), with more than 90 per cent of this growth in the developing world. These are similar figures to those produced by the World Bank in 1991 which were assumed in the IPCC IS 92a scenario.

All four cases assume that there will be significant environmental and economic pressures to achieve major improvements in energy efficiency compared to historic performance, although to different degrees within the various economic groupings of countries. One of them,

WEC scenarios

Case	A	B1	B	C
Name	High growth	Modified reference	Reference	Ecologically driven
Economic Growth % p.a.	High	Moderate	Moderate	Moderate
OECD	2.4	2.4	2.4	2.4
CEE/CIS	2.4	2.4	2.4	2.4
DCs	5.6	4.6	4.6	4.6
World	3.8	3.3	3.3	3.3
Energy intensity reduction % p.a.	High	Moderate	High	Very high
OECD	-1.8	-1.9	-1.9	-2.8
CEE/CIS	-1.7	-1.2	-2.1	-2.1
DCs	-1.3	-0.8	-1.7	-2.4
World	-1.6	-1.3	-1.9	-2.4
Technology transfer	High	Moderate	High	Very high
Institutional improvements (world)	High	Moderate	High	Very high
Possible total demand (Gtoe)	Very high 17.2	High 16.0	Moderate 13.4	Low 11.3

large influence on energy demand and growth. The other scenarios (B and B1), denoted reference cases, are based on moderate assumptions about economic growth.

Details of the scenarios to the year 2020 are shown in Figs 11.3 and 11.4. As can be seen from Fig. 11.3, it is only in the developed world that

there is potential for containing future energy demand. Population growth and the need for economic development in developing countries make it inevitable that they will, for many decades, consume increased amounts of energy. For all the scenarios fossil fuels dominate the energy mix over the next few decades

scenario C, assumes very strong pressure to reduce the emissions of greenhouse gases in order to combat global warming. Table 11.2 presents the detailed assumptions underlying the four scenarios.

Table 11.2 refers to the 'energy intensity' which is the ratio of energy use to Gross Domestic Product (GDP); it is a measure of energy efficiency. A demanding rate of reduction in energy intensity is assumed for all the scenarios; for case C, the ecologically driven scenario, the rate assumed is very

demanding indeed. The main difference between the modified reference case B1 and the reference case B is that, in B1, the rate of reduction of energy intensity assumed for the economies in transition is less than in case B and for the developing countries is only half that in case B.

With somewhat less detail, scenarios A, B and C have been extended to the year 2100. Global energy demand can be expected to continue to increase, but by that time the availability of fossil fuels will be

more limited and new renewables can be expected to contribute substantially to the energy mix for all scenarios. Some of the characteristics of the scenarios out to 2100 are listed in Table 11.3

The IPCC scenario IS 92a—the business-as-usual scenario—assumes moderate economic growth and no strong environmental pressures to cut emissions (more details in the caption to Table 11.3).

WEC scenarios to the year 2100

Case		A		B		C	
	1990	2050	2100	2050	2100	2050	2100
Global energy demand (Gtoe)	8.8	27	42	23	33	15	20
Fossil fuels (% of primary energy)	77	58	40	57	33	58	15
Nuclear (% of primary energy)	5	14	29	15	28	8	11
New renewables (% of primary energy)	2	15	24	14	26	20	50
Annual CO_2 emissions from fossil fuels	6.0	14.9	16.6	12.2	11.7	7.3	2.5
Annual CO_2 emissions from fossil fuels (% change on 1990)		152	181	107	98	24	-59

TABLE 11.2 Assumptions underlying the four WEC energy scenarios[7]. See glossary for explanation of abbreviations.

TABLE 11.3 Some characteristics of the WEC scenarios out to the year 2100[8]. For Scenario IS 92a, the comparable figures are: 40 Gtoe for the global energy demand in 2100 (about 53% supplied by fossil fuels) and 13.2 and 19.8 Gt respectively for the CO_2 emissions from fossil fuels in 2050 and 2100.

(Fig. 11.4). The contribution from nuclear power is assumed to grow in all the cases. New renewable energy sources play an increasing role, although apart from Case C their contribution is modest.

For the 'ecologically driven' WEC scenario C the energy demand

FIG. 11.3 **Primary energy demand by economic groupings of countries**[9].

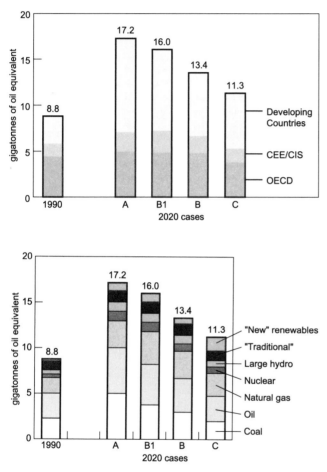

FIG. 11.4 **Primary energy supply mix**[10].

coming from new renewable energy sources ('modern' biomass, solar, wind and so on). A growth in energy supply from these new renewable sources from 2 per cent in 1990 to 12 per cent in 2020 (by when 1.4 Gtoe per year would be coming from these sources) is thought to be just about feasible if their development is given sufficient support; by the year 2100 under scenario C, 50 per cent of energy supply is assumed to come from these sources. The WEC report points out that 'cost effective research, development and installation involving financing which only governments can supply will be needed if these sources of energy are to be implemented on the large scale shown in the Ecologically Driven Case C'. Renewable energy sources will be discussed further later in the chapter.

It is interesting to compare the future projections generated by the WEC with the IPCC business-as-usual scenario IS 92a. Up to the year 2020 the energy demand and the carbon dioxide emissions (Fig. 3.5) for IS 92a are similar to those of the WEC Reference scenario B1. After about the middle of the century, the WEC scenarios assume more than the IPCC ones that growth in energy demand and also in carbon dioxide emissions will begin to be limited by the increasing scarcity and price of oil and gas.

For all the scenarios, except WEC scenario C, atmospheric concentrations of carbon dioxide (Fig. 3.6) continue to rise throughout next century. The WEC scenario C comes near to stabilizing emissions over the period to 2020 (see Figs 3.5 and 3.6) and to

in 2020 is about 30 per cent more than that in 1990 and 30 per cent less than that for scenario A. That scenario C, following the WEC Committee on Renewable Energy, assumes both that there will be large increases in efficiency leading to a reduced energy demand and also a substantial growth in the share of primary energy supply stabilizing atmospheric con-centrations by 2100.

The following sections address how increased energy conservation and efficiency can be achieved and what developments can be realized in new renewable energy sources. These are the means by which the world could come near to the achievement of WEC scenario C rather than scenarios A or B.

Energy conservation and efficiency

If we turn lights off in our homes when we do not need them or if we turn down the thermostat by a degree or two so that we are less warm, we are conserving or indeed saving energy. But are such actions significant in overall energy terms? Is it realistic to plan for really worthwhile savings in our use of energy?

To illustrate what might be possible, I will consider the efficiency with which energy is currently used. The energy available in the coal, oil, gas, uranium, hydraulic or wind power is *primary energy*. It is rarely used directly, but is transformed into light, useful heat, motor power and so on. The process of energy conversion, transmission and transformation into its final useful form involves a proportion of the heat being wasted. To provide 1 unit of electrical power at the point of use typically requires

Efficiency of appliances

In Table 11.4, figures are given for the average annual electricity consumption of typical domestic appliances used in a home in the United Kingdom. If, in replacing appliances, everyone bought the most efficient available, their total electricity consumption could easily drop by more than half.

Savings in electricity also deliver a net economic return to the user. Take for example fluorescent light bulbs which are as bright as ordinary light bulbs, but use a quarter of the electricity and last eight times as long before they have to be replaced. A 15 watt compact fluorescent bulb (equivalent to a 60 watt ordinary incandescent light bulb) costing £10 will use about £10 worth of electricity over its lifetime. To cover the same period eight ordinary bulbs would be needed costing about £4 but using £40 worth of electricity. The net saving is therefore about £24. Similar calculations could be carried out for the other appliances.

TABLE 11.4 **Annual appliance electricity consumption in kWh[11].**

Annual appliance electricity consumption in kWh per year

(kWh/year)	1990 Existing stock (UK)	Average new model (UK)	Best available model (worldwide)
Cooker	840	780	370
Washing machine	210	180	70
Dishwasher	500	430	300
Refrigerator	350	300	60
Fridge/Freezer	730	500	275
TV	200	140	100
Lighting	370	370	105

Insulation of buildings

About one and a half thousand million people live in cold climates where some heating in buildings is required. In most countries the energy demand of space heating in buildings is far greater than it need be if the buildings were better insulated.

Table 11.5 provides as an example details of two houses, showing that the provision of insulation in the roof, the walls and the windows can easily lead to the energy requirement for space heating being more than halved (from 5.8kW to 2.65kW). The cost of the insulation is small and is quickly recovered through the lower energy cost.

TABLE 11.5 **Two assumptions (one poorly insulated, and one well insulated) regarding construction of a detached, two-storey house with ground floor of size 8m x 8m, and the accompanying heat losses (U-values express the heat conduction of different components in watts per square metre per °C.)**

	Poorly insulated	**Well insulated**
Walls 150m² total area	Brick + cavity + block U-value 0.7	Brick + cavity + block with insulation in cavity of 75mm thickness U-value 0.3
Roof 64m² area	Uninsulated U-value 2.0	Covered with insulation of thickness 150mm U-value 0.2
Floor 64m² area	Uninsulated U-value 1.0	Includes insulation of thickness 50mm U-value 0.3
Windows 12m² total area	Single glazing U-value 5.7	Double glazing with low emissivity coating U-value 2.0
Heat losses (in kW) with 10°C temperature difference from inside to outside	roof 1.7 walls 1.1 · windows 0.7 floor 0.7	roof 0.2 walls 0.45 · windows 0.2 floor 0.2
Total heat loss (kilowatts)	4.2	1.05
Add heat (in kilowatts) needed for air changes (1.5 per hour)	1.60	1.60
Total heating required (kilowatts)	5.8	2.65

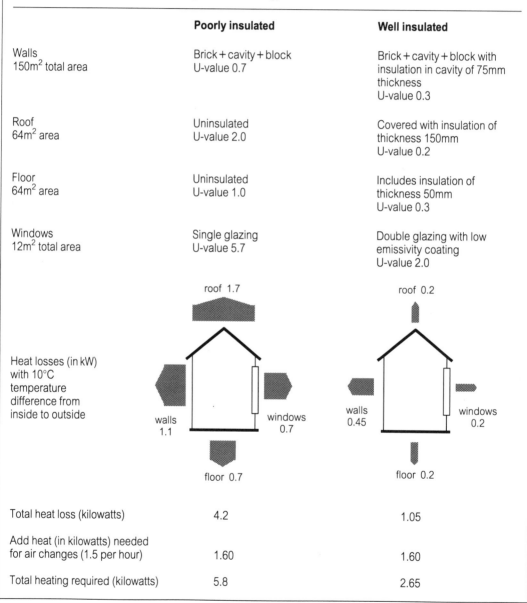

nearly 3 units of primary energy. An incandescent light bulb is about 3 per cent efficient in converting primary energy into light energy; unnecessary use of lighting reduces the overall efficiency to perhaps no more than 1 per cent[12]. Assessments have been carried out across all energy uses comparing actual energy use with that which would be consumed by ideal devices providing the same services. Although there is some difficulty in defining precisely the performance of such 'ideal' devices, assessments of this kind come up with world average end-use energy efficiencies of the order of 3 per cent. That sort of figure suggests that there is a large amount of room for improvement in energy efficiency, perhaps by at least three-fold[13]. The following paragraphs consider three areas where there are real possibilities of savings: in buildings, in industry and in transport.

First of all, what about energy in buildings? So that we are comfortable in them we heat them in winter and we cool them in summer. In the United States, for instance, about 36 per cent of the total use of energy is in buildings (about two-thirds of this in electricity), including about 20 per cent for their heating (including water heating) and about 3 per cent for cooling them[14].

Two ways in which really substantial energy savings can be made in buildings are by improving their insulation and by improving the efficiency of appliances. Many countries, including the UK and the USA still have relatively poor standards of building insulation compared, for instance, with Scandanavian countries. There are also large possibilities for the improvement of the efficiency of

appliances at relatively small cost (see box on p 157).

The results of a study in the United States have identified some of the large savings which could be made in the electricity used in buildings. The cost of such action would be less than the cost of the energy savings; overall therefore there would be a substantial net saving (Fig. 11.5). The twelve options in Fig. 11.5 together cover about 45 per cent of the amount of electricity used in residential buildings in the USA, which in 1989 was 1630 TWh or about 10 per cent of the USA's total energy use. The four options which provide the largest savings (together adding up to 60 per cent of the savings) are in the areas of commercial lighting, commercial air conditioning, residential appliances and residential space heating. Electricity companies in some parts

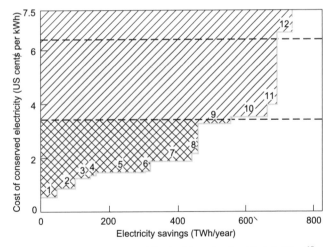

FIG. 11.5 Cost of various options (at 1989 prices) for saving electricity in buildings[15]. If the cost of conservation is less than the cost of the electricity saved, a net saving results. The various options are: (1) use of white surfaces to reduce the need for air conditioning; (2) residential lighting; (3) residential water heating; (4) commercial water heating; (5) commercial lighting; (6) commercial cooking; (7) commercial cooling; (8) commercial refrigeration; (9) residential appliances; (10) residential space heating; (11) commercial and industrial space heating; (12) commercial ventilation. The shaded areas are below 7.5¢ per kWh (the all-sector average electricity price) and 3.5¢ per kWh (typical operating cost of US electricity generation).

of the United States are contracting to implement some of these energy saving measures as an alternative to the installation of new capacity—at significant profit both to the companies and its customers. Similar savings would be possible in other developed countries. Major savings at least as large in percentage terms could also be made in countries with economies in transition and in developing countries if existing plant and equipment were used more efficiently.

Similar room for efficiency savings exists in industry. The installation of relatively simple control technology often provides large potential for energy reduction at a substantial net saving in cost.

The growth of motor transport

Since the 1950s there has been a phenomenal growth in road transport (Fig. 11.6) especially in the industrialized Western countries. In these countries road transport accounts for over 80 per cent of all transport and also for nearly 50 per cent of total oil consumption.

The potential for future growth is not only illustrated by the trend shown in Fig. 11.6, but also by the enormous disparity in car ownership and use between different countries (Table 11.6).

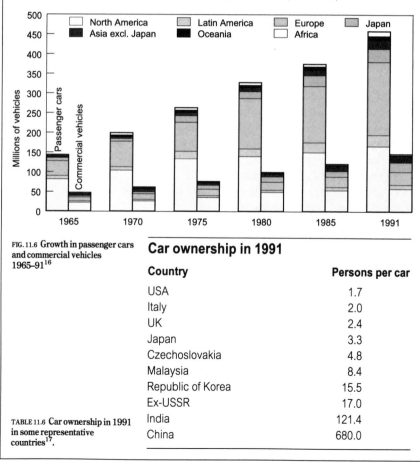

FIG. 11.6 Growth in passenger cars and commercial vehicles 1965–91[16]

Car ownership in 1991

Country	Persons per car
USA	1.7
Italy	2.0
UK	2.4
Japan	3.3
Czechoslovakia	4.8
Malaysia	8.4
Republic of Korea	15.5
Ex-USSR	17.0
India	121.4
China	680.0

TABLE 11.6 Car ownership in 1991 in some representative countries[17].

The cogeneration of heat and power, which already enables electricity generators to make better use of heat which would otherwise be wasted, is particularly applicable to some industrial plants where large amounts of both heat and power can be required. To take an example: British Sugar with an annual turnover of £700 million now spends about £21 million on energy. Since 1980 the energy consumed per tonne of sugar has been reduced by 41 per cent through low grade heat recovery, cogeneration schemes and better control of heating and lighting[18]. Many studies in industrialized countries indicate that savings of 30 per cent or more could be made in the industrial sector at a net saving in overall economic terms[19].

In the developed world between one fifth and one quarter of energy use is for transport, and it is a rapidly growing proportion. Road transport accounts for the largest proportion of this (see box), over 80 per cent in industrialized countries; air transport is next at 13 per cent. In the United Kingdom, for instance, if the present trend in motor-car use continues, in about thirty years there will be twice as many cars doing twice as many miles—and the situation is very similar elsewhere. The proportional growth in developing countries could be even more rapid. The advantages conferred by the motor-car, the convenience, freedom and flexibility which it brings, mean that growth in its use is bound to continue. Increased prosperity also brings with it increased movement of freight.

There are two types of action which can be taken to curb the energy use of transport. The first is to increase the efficiency of fuel use.

We cannot expect the average car to compete with the vehicle which, in 1992, set a record by covering over 12,000km on one gallon of petrol—a journey which serves to illustrate how inefficiently we use energy for transport! But technology is available now through which the average fuel consumption of the current fleet of motor cars could be halved while maintaining an adequate performance. The second action is to plan cities and other developments so as to lessen the need for transport and to make personalized transport less necessary—work, leisure and shopping should all be easily accessible by public transport, or by walking or cycling. Such planning is already being taken seriously in some countries where the need to reverse some current trends (for instance the growth of large out-of-town shopping developments) is also recognized.

Before leaving the question of energy efficiency, two other areas should be mentioned. First, there are the questions of how far the efficiency of large power stations or other installations burning fossil fuels can be improved and also of whether their carbon dioxide emissions could be prevented from entering the atmosphere.

The efficiency of coal-fired power stations, for instance, has improved from about 32 per cent, a typical value of 20 years ago, to about 36 per cent for a modern fluidized bed plant of today. Such an improvement is significant in environmental terms and it is important that means be provided for the latest, most efficient technology to be available and atttractive to rapidly industrializing countries such as China. Substantial

161

further gains in overall efficiency can be realized by making sure that the large quantities of low-grade heat generated by power stations is not wasted but utilised, for instance in combined heat and power (CHP) schemes.

Regarding carbon dioxide removal from the flue gases, it must be realized that the amounts are very large; the gas to be removed weighs thousands of millions of tons. A number of proposals for such removal have been proposed, for instance pumping the carbon dioxide into spent oil or gas wells or into the deep ocean. However, even in the most favourable circumstances (for instance when power stations are close to oil or gas fields), the cost of removal would be significant (perhaps 40 per cent on top of the energy cost) and for deposition in the deep ocean would be large (perhaps 100 per cent on top of the energy cost)[21]. Further, if deposition in the deep ocean is seriously contemplated, the potential

hazards associated with it need to be very carefully assessed.

The other area needing brief consideration is that of the 10 per cent or so of world energy which is not commercial and which comes from traditional sources, largely from wood or other biomass. This is the only source of energy for half the world's population. Although all of these sources are renewable, it is still important that they are employed efficiently, and a great deal of room for increased efficiency exists. For instance, much cooking is still carried out on open fires where only about 5 per cent of the heat reaches the inside of the cooking pot. The introduction of a simple stove can increase this to 20 per cent or with a little elaboration to 50 per cent[22]. There is often considerable consumer resistance to the introduction of such stoves, but if they were widely used, the saving worldwide in fuelwood would be enormous. Other means of reducing fuelwood demand would be to

TABLE 11.7 **Contributions to world energy supply** (in millions of tonnes of oil equivalent) from renewable sources in 1990 and as assumed under the WEC scenario C in 2020[20].

Renewable energy sources

	1990 Mtoe	1990 % of World Energy	2020 Mtoe	2020 % of World Energy
'Modern' Biomass	121	1.4	561	5.0
Solar	12	0.1	355	3.1
Wind	1	0.0	215	1.9
Geothermal	12	0.1	91	0.8
'Small' hydro	18	0.2	69	0.6
Oceanic	0	0.0	54	0.5
Total (new renewable sources)	164	1.8	1345	11.9
'Large' hydro	465	5.3	661	5.8
'Traditional' biomass	930	10.6	1060	9.3
Total (all renewables)	1559	17.7	3093	27.0

encourage alternatives such as the use of fuel from crop wastes, of methane from sewage or other waste material or of solar cookers (mentioned again later on).

Renewable energy

To put our energy use in context it is interesting to realize that the energy incident on the Earth from the sun amounts to about 180 thousand million million watts (or 18,000 Terawatts, $1TW = 10^{12}W$). This is about 15,000 times the world's average energy use of about 12 million million watts (12 TW). As much energy arrives at the Earth from the sun in 40 minutes as we use in a whole year. So, providing we can harness it satisfactorily and economically, there is plenty of renewable energy coming in from the sun to provide for all the demands human society can conceivably make.

There are various ways in which solar energy is converted into forms that we can use; it is interesting to look at the efficiencies of these conversions. If the solar energy is concentrated, by mirrors for instance,

almost all of it can be made available as heat energy. Between 1 and 2 per cent of solar energy is converted through atmospheric circulation into wind energy, which although concentrated in windy places is still distributed through the whole atmosphere. About 20 per cent of solar energy is used in evaporating water from the Earth's surface which eventually falls as precipitation, giving the possibility of hydro-power. Living material turns sunlight into energy through photosynthesis with an efficiency of around 1 per cent for the best crops. Finally, photovoltaic (PV) cells convert sunlight into electricity with an efficiency which for the best modern cells can be over 20 per cent.

Around the year 1900, very early on in the production of commercial electricity, water power was an obvious source and from the beginning made an important contribution. Hydroelectric schemes now supply about 6 per cent of the world's commercial energy. Other renewable sources of commercial energy have, however, been slow to take off. In 1990, only about

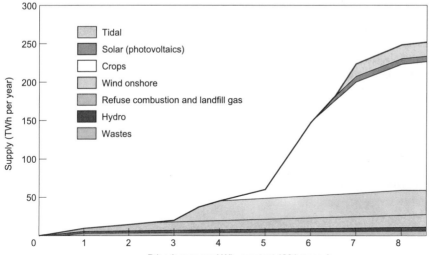

FIG. 11.7 **Estimated contribution to electricity generation in the UK from renewable energy sources in the year 2025 as a function of the electricity price with an assumption of an 8% discount rate. The assessment for electricity generated from crops is considered a preliminary one[24].**

2 per cent of the world's commercial energy came from all renewable sources other than large hydro[23]. Of this 2 per cent, about three-quarters was from 'modern' biomass, the other 0.5 per cent being shared between solar, geothermal, wind energy and small hydro sources.

Returning to commercial energy generation, in order to put renewable sources into context, it is useful to inspect the detailed projection of the WEC (Table 11.7) for the contributions from different 'new renewable' sources which make up the 12 per cent of total energy supply in the year 2020 assumed for the WEC scenario C. The main growth expected is in energy from 'modern' biomass and from solar and wind energy sources.

In the following paragraphs, the main renewable sources are described in turn and their possibilities for growth considered[25]. Most of them are employed for the production of electricity; in the case of biomass, liquid or gaseous fuels can also be produced.

Hydro-power

Hydro-power, the oldest form of renewable energy, is well established and is competitive economically with electricity generated by other means. Some hydroelectric schemes are extremely large. Two of the world's largest schemes, each of over 10,000MW capacity, are in South America at Guri in Venezuela and at Itaipu on the borders of Brazil and Paraguay. It is estimated[26] that there is potential for further exploitation of hydroelectric capacity to three or four times what has currently been developed, much of this undeveloped potential being in the former Soviet Union and in developing countries.

Large schemes, however, can have significant social impact (such as the movement of population from the reservoir site), environmental consequences (for example, loss of land, of species and of sedimentation to the lower reaches of the river), and problems of their own such as silting up, which have to be thoroughly addressed before they can be undertaken.

But hydroelectric schemes do not have to be large; Table 11.7 distinguishes between large and small hydroelectric sources. Many units exist generating a few kilowatts only which may supply one farm or a small village. The attractiveness of small schemes is that they provide a locally based supply at modest cost. For instance, in China there are a large number of small hydro plants of less than 10MW which aggregated in 1990 to some 4,000MW in total capacity; a further 2,000MW of such schemes are currently under development in that country. Many more possibilities exist for the exploitation of the potential of small rivers and streams in many parts of the world. About 1,000MW capacity per year is currently being installed globally in small hydro schemes. Given appropriate incentives by governments and the availability of finance it is estimated that this rate of installation could double[27].

An important facility provided by some hydro schemes is that of pumped storage. Using surplus electricity available in off-peak hours, water can be pumped from a lower reservoir to a higher one. Then, at other times, by reversing the process, electricity can be generated to meet periods of peak demand. The efficiency of conversion can be as high as 80 per cent and the response time a few seconds, so reducing the need to keep other generating capacity in reserve. In 1990 about 75 GW of pumped storage capacity was available worldwide with a further 25 GW under construction[28].

Biomass as fuel

Second in current importance as a renewable energy source is the use of biomass as a fuel. The word biomass in this context covers domestic, industrial and agricultural dry waste material, wet waste material and crops, all of which can be used as fuel to power electricity generators and some of which are appropriate to use for the manufacture of liquid or gaseous fuels.

There is considerable public awareness of the vast amount of waste produced in modern society. The United Kingdom, for example, generates each year somewhat over 30 million tonnes of domestic solid waste, or about half a ton for every citizen; this is a typical value for a country in the developed world. Even with major programmes for recycling some of it, large quantities would still remain. If it were all incinerated for power generation about 1.7 GW could be generated, about 5 per cent of the UK's electricity requirement[29].

But what about the greenhouse gas generation from waste incineration? Carbon dioxide is of course produced from it, which

Sugar cane as biomass

Sugar cane production yields two kinds of biomass fuel suitable for gasification (Fig. 11.8), known as bagasse and barbojo. Bagasse is the residue from crushing the cane and is thus available during the milling season; barbojo consists of the tops and leaves of the cane plant that could be stored for use after the milling season. It has been estimated that using these sugar cane sources, within thirty years or so, the 80 sugar cane producing countries in the developing world could generate two-thirds of their current electricity needs at a price competitive with fossil fuel energy sources[30].

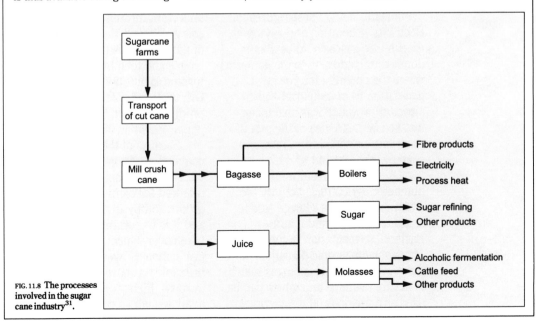

FIG. 11.8 The processes involved in the sugar cane industry[31].

165

contributes to the greenhouse effect. However, the alternative method of disposal is landfill (most of the waste in the UK currently is disposed of that way). Decay of the waste over time produces carbon dioxide and methane in roughly equal quantities. Some of the methane can be collected and used as a fuel for power generation. However, only a fraction of it can be captured; the rest leaks away. Because methane is a much more effective greenhouse gas than carbon dioxide, the leaked methane makes a substantial contribution to the greenhouse effect. Detailed calculations show that if all United Kingdom domestic waste were incinerated for power generation rather than landfilled, the net saving per year in greenhouse gas emissions would be equivalent to about 10 million tons of carbon as carbon dioxide[32]. Since this is about 5 per cent of the total United Kingdom greenhouse gas emissions, we can infer that power generation from waste could be a significant contribution to the reduction in overall emissions. These figures for electricity generation and saving in greenhouse gases could be about doubled if in addition to the domestic waste the potential for power generation from industrial and agricultural waste was also taken into account. An idea of the cost of electricity provided from these sources is shown in Fig. 11.7.

Other wastes resulting from human or agricultural activity are wet wastes such as sewage sludge and farm slurries and manures. Bacterial fermentation in the absence of oxygen (anaerobic digestion) of these wastes produces biogas which is mostly methane and which can be used as a fuel to produce energy.

Although not a large potential energy source (Fig. 11.7), there is room for its contribution to increase.

The third potential biomass contribution is from the use of crops as a fuel. Here the potential is large. It is a genuinely renewable resource in that the carbon dioxide which is emitted when the biomass is burnt is turned back again into carbon, through the process of photosynthesis, in the renewed biomass when it is grown again. Many different crops can be employed as biomass for energy production. In Brazil, for instance, since the 1970s large plantations of sugar cane have produced alcohol for use as a fuel mainly in transport, generating, incidentally, much less local pollution than petrol or diesel fuel from fossil sources. A lot of potential has been recognized for the sugar cane industry to produce both sugar and energy together with other byproducts as well (see box on p 165). Biomass from wood plantations on agricultural land no longer needed for food crops features as an important future source in Sweden's energy plans[33]; the most efficient use of the biomass is first to turn it into biogas and then burn it in a gas turbine to produce electricity. For the UK, trials indicate that the most promising option is willow and poplar grown in coppices[34].

Because of the low efficiency of conversion of solar energy to biomass, the amount of land required for significant energy production by this means is large— and it is important that land is not taken over which is required for food production. However, there is in principle no shortage of land for this purpose. Plenty of suitable crops are available which could be grown on

166

land only marginally useful for agriculture. The main drawback is the cost of energy production from biomass (Fig. 11.7). The basic cost of electricity generated this way is currently about twice that generated from fossil fuel sources, and the cost of alcohol from biomass as a fuel for transport is about twice the basic cost of petrol or diesel. However, in some countries the expenditure which would otherwise be incurred in the setting aside of land from agricultural use for food production could be set against that extra cost.

Wind energy

Energy from the wind is not new. Two hundred years ago windmills were a common feature of the European landscape; for example, in 1800 there were over 10,000 working windmills in Britain. During the past few years they have again become familiar on the skyline, for instance, in south-west England and Wales. Slim, tall, sleek objects silhouetted against the sky, they do not have the rustic elegance of the old windmills, but they are much more efficient. A typical wind energy generator will have a two or three-bladed propeller 33m in diameter and a rate of power generation in a wind speed of 12m/sec (43km/hr, 27mph or Beaufort Force 6), of 300 kilowatts. On a site with an average wind speed of about 7.5m/sec (an average value for the west of Great Britain) it will generate an average power of about 100 kW. The generators are often sited close to each other in wind farms which may include several dozen such devices.

From the point of view of the electricity generating companies the difficulty with the generation of

electrical power from wind is that it is intermittent. There are substantial periods with no generation at all. The generating companies can cope with this in the context of a national electricity grid which pools electrical power from different sources providing that the proportion from wind sources is not too large.

The power generated from the wind depends on the cube of the wind speed: a wind speed of 12.5m/sec is twice as effective as one of 10m/sec. It makes sense therefore to mount wind farms on the windiest sites available. The west coasts of the British Isles have some of the strongest winds in Europe and are therefore particularly suitable. It has been estimated that wind energy could within reasonable cost (Fig. 11.7) contribute up to 10 per cent of the UK's electricity supply[35]. Such a provision would require more than 40,000 330 kW wind generators; these would need to be carefully sited

---

Wind power on Fair Isle

A good example[36] of a site where wind power has been put to good effect is Fair Isle, an isolated island in the North Sea north of the Scottish mainland. Until recently, the population of 70 depended on coal and oil for heat, petrol for vehicles and diesel for electricity generation. A 50 kW wind generator was installed in 1982 to generate electricity from the persistent strong winds of average speed over 8 m/sec (29km/h or 18 mph). The electricity is available for a wide variety of purposes; at a relatively high price for lighting and electronic devices and at a lower price controlled amounts are available (wind permitting) for comfort heat and water heating. At the frequent periods of excessive wind further heat is available for heating glasshouses and a small swimming pool. An electric vehicle has been charged from the system to illustrate a further use for the energy.

With the installation of the wind generator, which now supplies over 90 per cent of the island's electricity, electricity consumption has risen about fourfold and the average electricity costs have fallen from 13p/kWh to 4p/kWh.

Solar water heating

The essential components of a solar water heater (Fig. 11.9) are a set of tubes in which the water flows embedded in a black plate insulated from behind and covered with a glass plate on the side facing the sun. A storage tank for the hot water is also required. A more efficient (though more expensive) design is to surround the black tubes with a vacuum to provide more complete insulation.

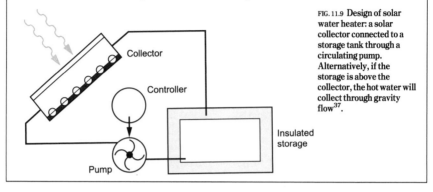

FIG. 11.9 **Design of solar water heater: a solar collector connected to a storage tank through a circulating pump. Alternatively, if the storage is above the collector, the hot water will collect through gravity flow[37].**

Solar energy in building design

All buildings benefit from unplanned gains of solar energy through windows and, to a lesser extent, through the warming of walls and roofs. This is called 'passive solar gain'; for a typical house in the United Kingdom it will contribute about 15 per cent of the annual space heating requirements. With 'passive solar design' this can relatively easily and inexpensively be increased to around 30 per cent, while increasing the overall degree of comfort and amenity. The main features of such design are to place, so far as is possible, the principal living rooms with their large windows on the south side of the house, with the cooler areas such as corridors, stairs, cupboards and garages with the minimum of window area arranged to provide a buffer on the north side. Conservatories can also be strategically placed to trap some solar heat in the winter.

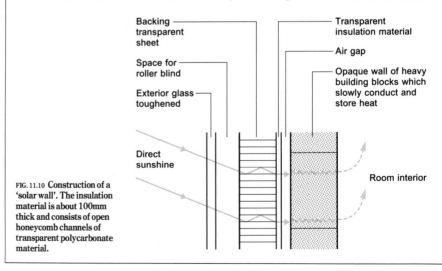

FIG. 11.10 **Construction of a 'solar wall'. The insulation material is about 100mm thick and consists of open honeycomb channels of transparent polycarbonate material.**

if visual amenity were not to be lost. It would also require that the electricity grid be reinforced so that power generated in the Hebrides islands off the coast of Scotland can be transmitted south to England. In the longer term this provision could perhaps be doubled, especially if an effective means for energy storage (for instance using hydrogen; more of that possibility later in the chapter) were developed. Many other countries with windy coast lines or mountain regions could also make substantial use of wind energy.

Wind energy is also particularly suitable for the generation of electricity at isolated sites to which the transmission costs of electricity

from other sources would be unacceptable. Because of the wind's intermittency, some storage of electricity or some back-up means of generation has to be provided as well. The installation on Fair Isle (see box) is a good example of an efficient and versatile system. Small wind turbines also provide an ideal means for charging batteries in isolated locations; for instance, about 83,000 are in use by Mongolian herdsmen. Wind energy is often also an ideal source for water pumps.

With the large growth during the last few years in the installations of wind generators, economies of scale have brought down the cost of the electricity generated. Eventually it can be expected to approach the cost of electricity generated from fossil fuels (Fig. 11.7).

Energy from the sun

The simplest way of making use of energy from the sun is to turn it into heat. A black surface directly facing full sunlight can absorb about 1 kilowatt for each square metre of surface. In countries with a high incidence of sunshine it is a effective and cheap means providing domestic hot water (see box), which is extensively employed in countries such as Australia, Israel, Japan and the southern states of the United States of America. In tropical countries, a solar cooking stove can provide an efficient alternative to stoves burning wood and other traditional fuels. Thermal energy from the sun can also be employed effectively in buildings, in order to provide a modest boost towards heating the building in winter and, more importantly, to provide for a greater degree of comfort and a more pleasant environment (see box).

The wall of a building can be designed specifically to act as a passive solar collector, in which case it is known as 'solar wall' (Fig. 11.10)[38]. Its construction enables sunlight, after passing through an insulating layer, to heat the surface of a wall of heavy building blocks which retain the heat and slowly conduct it into the building. The insulating layer, although allowing sunlight to pass through, prevents thermal radiation from passing out. A retractable reflective blind can be placed in front of the insulation at night or during the summer when heating of the building is not required. A set of student residences for 376 students at Strathclyde University in Glasgow has been built with a 'solar wall' on its south-facing side. Even under the comparatively unfavourable conditions during winter in Glasgow (the average duration of bright sunshine in January is only just over one hour per day) there is a significant net gain of heat through the wall to the building.

Solar heat can also be employed to provide heating to produce steam for the generation of electricity. To produce significant quantities of steam, the solar energy has to be concentrated by using mirrors. One arrangement employs trough-shaped mirrors aligned east-west which focus the sun on to an insulated black absorbing tube running the length of the mirror. A number of such installations have been built, particularly in the United States, where solar thermal installations provide over 350 MW of commercial electricity. The high capital cost of such installations, however, assuming a reasonable pay-back period, translates into an electricity cost which, at the moment, is at least three times that from most conventional sources.

Sunlight can be converted directly into electricity by means of photo-voltaic (PV) solar cells (see box). They appear in a variety of ways in everyday life; as power sources, for instance for small calculators or watches. Spacecraft are covered with 'solar panels' or possess 'solar arrays' covered with PV cells providing electrical power for the spacecraft. Their efficiency for

conversion of solar energy into electrical energy is now generally between just under 10 per cent and 20 per cent. A panel of cells of area one square metre facing full sunlight will therefore deliver between 100 and 200 watts of electrical power. The cost of energy from solar cells has reduced dramatically over the past twenty years (see box); so much so that they can now be employed for a wide range of applications and can also begin to contribute to the large-scale generation of electricity.

Small PV installations are eminently suitable, especially for developing countries, to provide local sources of electricity in rural areas. About a third of the world's population have no access to electricity from a central source. Their predominant need is for small amounts of power for lighting, for radio and television, for refrigerators (for example, for vaccines at a health clinic) and for pumping water. The cost of PV installations for these purposes is now competitive with other means of generation (such as diesel units) and around 100,000 DLKs (Domestic Lighting and broadcast reception charging Kits— Fig. 11.11) have now been installed worldwide[40]. Larger installations are required for public buildings such as hospitals. A small hospital, again in Sri Lanka, through assistance from the Australian government, has installed a 1.3kW solar array, backed up by a 2,200 amp-hour battery, to provide for lighting, refrigeration for vaccines, autoclave sterilization, pumping for hot water (produced through a solar thermal system) and radio.

In 1990 the world capacity for production of solar cells was equivalent to a peak electrical energy

FIG. 11.11 A simple solar power installation for a small home is being marketed in Sri Lanka for less than $(US)100[39]. An array of 36 solar cells, each 10cm in diameter, provides 40 watts of peak power. This is sufficient to charge a car battery which can power up to three 9 watt fluorescent lights and three hours of radio and one hour of television each day. With more restricted use of these devices or with a larger solar array, a small refrigerator can be added to the system.

Local solar energy supply

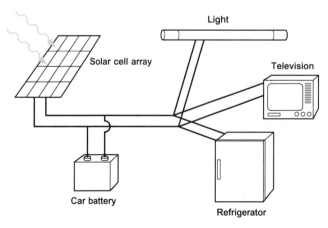

170

of about 50MW. To meet the estimates under the WEC scenario C of their projected contribution to world energy supply of at least 100 GW[41] by the year 2020 (Table 11.7), production will need to increase by over a factor of a thousand. In the short term, increased development of local installations is likely to have priority; after the turn of the century, with the expectation of a significant cost reduction (Fig. 11.12), penetration into large-scale electricity generation will become more possible.

Other renewable energies

We have so far covered the renewable energy sources for which there is potential for growth on a scale which can make a significant contribution to overall world energy demand. We should also mention briefly other renewable energy technologies which contribute to global energy production and which are of particular importance in certain regions, namely geothermal energy from deep in the ground and energies from the tides, waves or thermal gradients in the ocean.

The photovoltaic solar cell

The silicon photovoltaic (PV) solar cell consists of a thin slice of silicon into which appropriate impurities have been introduced to create what is known as a p-n junction. The most efficient cells are sophisticated constructions using crystalline silicon as the basic material; they possess efficiencies for the conversion of solar energy into electricity typically of 15-20 per cent; experimental cells have been produced with efficiencies well over 20 per cent[42].

Single crystal silicon is not convenient for mass production, but amorphous silicon, with a conversion efficiency of around 10 per cent, can be deposited in a continuous process on to thin films[43]. Other alloys (such as cadmium telluride and copper indium selenide) with similar photovoltaic properties can also be deposited in this way and, because they have higher efficiencies than amorphous silicon, may become important in the future[44].

Cost is of critical importance if PV solar cells are going to make a significant contribution to energy supply. This has been coming down rapidly. More efficient methods and larger-scale production are bringing the cost of solar electricity down to levels where it can compete with other sources. Projections up to the year 2020 of the likely cost of electricity generated from PV sources are compared with that from conventional sources in Fig. 11.12.

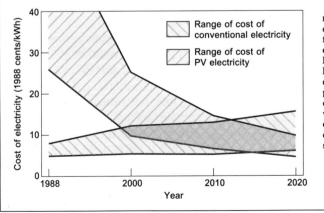

FIG. 11.12 **The falling cost of electricity produced from PV solar cells as estimated by the World Energy Council**[45]. **Projections of the range of costs of the production of PV electricity is compared with the range of cost of electricity from conventional fossil fuel sources.**

171

The presence of geothermal energy from deep down in the Earth's crust makes itself apparent in volcanic eruptions and less dramatically in geysers and hot springs. The energy available in favourable locations may be employed directly for heating purposes or for generating electrical power. Although very important in particular places, for instance in Iceland, it is currently only a small contributor (between 0.1 and 0.2 per cent) to total world energy; its contribution could rise to the order of 1 per cent during the next few decades (Table 11.7).

Large amounts of energy are in principle available in the movements and the temperature gradients of the ocean; but they are in general not easy to exploit. Tidal energy is the only one currently contributing significantly to commercial energy production. The largest tidal energy installation is at La Rance in France; it has a capacity of 240MW. Several estuaries in the world have been extensively studied as potential sites for tidal energy installations. The Severn Estuary in the United Kingdom, for instance, has the

potential to generate a peak power of over 8,000MW or about 6 per cent of the total UK electricity demand. Although the cost of the electricity generated from the largest schemes could be competitive, the main deterrents to such schemes are the high capital up-front cost and the significant environmental impacts that can be associated with them.

The financing of renewable energy

Renewable energy on the scale envisaged by WEC scenario C will only be realized if it is seen to be competitive in cost with energy from other sources and if the investment it requires is made available. Considering electricity and the cost of various sources of its supply, in Fig. 11.13 comparison is made between these costs and the carbon dioxide emissions from the different sources. Under some circumstances renewable energy sources are already competitive in cost, for instance in providing local sources of energy where the cost of transporting electricity or other fuel would be significant; some examples of this (such as Fair Isle in Scotland) have been given. However, if renewables are to begin to displace fossil fuels to the extent required by scenario C, there will need to be financial arrangements to bring about the change.

One way of engineering such arrangements would be to allocate an environmental cost to carbon dioxide emissions. Supposing, for instance, that an environmental cost of between $50 and $100 per tonne of carbon (figures mentioned in this context at the end of Chapter 9) were to be associated with carbon dioxide emissions, between 1 and 2 cents per kWh would be added to the price of

FIG. 11.13 Illustrating the cost-effectiveness of electricity supply options: the cost of electricity generation in the United States from various sources plotted against their generation of carbon dioxide emissions (expressed in terms of kg of carbon as carbon dioxide). Also plotted is the figure representing the average for generation in the United States in 1990 (the US mix). Distinction is made between coal generating plants which have fully depreciated and new pulverized coal plants. A range of values is given for nuclear energy and for renewable energy sources[46].

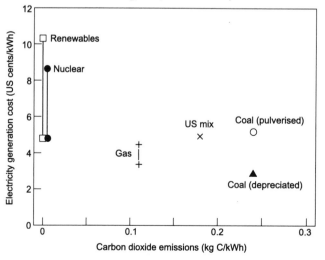

electricity from fossil fuel sources (Fig. 11.13)—which could begin to bring some renewables (for instance, biomass and wind energy) into competition with them.

Electricity, however, only accounts for about one third of the world's primary energy use. What about the possibility of renewable energy replacing solid and liquid fuels used for heating, for industry and for transport? It has already been mentioned that currently, liquid fuels such as ethanol derived from biomass are about twice as expensive as those derived from oil. Although there is an expectation that the processing of biomass will become more efficient[47]—the rapid development of technologies in bioengineering will help—it is unlikely that in the short term the substitution of biomass-derived fuels will occur on a large scale without the application of appropriate financial incentives. Arrangements of a similar kind to those we have already mentioned for electricity production would need to be put into place.

Looking at the investment required, annual investment in the world's energy industry is already large; it currently runs at between 3 and 4 per cent of GWP. The WEC estimates that cumulative world investment in energy supply will continue at least at the same level and up to the year 2020 will approach 30 million million US dollars at 1992 prices. They also estimate that to achieve the WEC scenario C, investment of at least 2.4 million million US dollars in new renewable energy sources would be required over the same period. Although this latter sum is less than 10 per cent of the world's total investment in energy, appropriate encouragement

and incentives will be required if it is to be realized.

Technology for the longer term

This chapter has concentrated mostly on what can be achieved with available and proven technology during the next few decades. It is also interesting to speculate about the more distant future and what relatively new technologies may become dominant later next century. In doing so, of course, we are almost certainly going to paint a more conservative picture than will actually occur. Imagine how well we would have done if asked in 1890 to speculate about technology change by the 1980s! Technology will certainly surprise us with possibilities not thought of at the moment. But that need not deter us from being speculative!

Of the renewable energy sources mentioned above, the direct conversion of sunlight into electricity by photovoltaic (PV) cells is in many ways the most attractive. There are many regions of the world where sunshine is plentiful and where suitable land not useful for other purposes would be readily available. It is a very clean, non-polluting

FIG. 11.14 A solar photo-voltaic (PV) electrolytic hydrogen system[48].

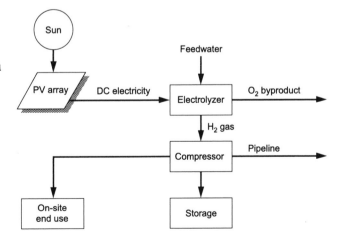

technology, easily adaptable to mass production. The cost of PV electricity has been coming down rapidly—a trend which is likely to continue especially with an increased scale of production.

But there are problems. Power from PV installations is only available during the day and such installations will usually be remote from places where the electricity is to be used. Means of energy storage and transport are therefore required. The best suggestion here is for hydrogen gas to be used as a storage medium. It can be generated directly from the PV electricity by the electrolysis of water (Fig. 11.14). This can be a very efficient process; over 90 per cent of the electrical energy can be stored in the hydrogen. The hydrogen can then be transported by pipeline or by bulk transport. Again, hydrogen is an extremely non-polluting energy source, which can easily be applied to most of the uses for which energy is required. It is perhaps particularly attractive for vehicle transport as, by means of a fuel cell, the energy in the hydrogen can be turned back into electricity for driving the vehicle, again rather efficiently and cleanly.

All the technology necessary for a solar-hydrogen energy economy is available now, although the cost of energy supplied this way would at the moment be several times that from fossil fuel sources[49]. As the technology for its further development progresses, the cost will undoubtedly reduce substantially. If its attractiveness from an environmental point of view were recognized as a dominant reason for its rapid development, the solar-hydrogen economy could take off more rapidly than most energy analysts are currently predicting.

Nuclear energy

An energy source mentioned rather little so far is nuclear energy. It is not strictly a renewable source, but it has considerable attractiveness from the point of view of sustainable development because it does not produce greenhouse gas emissions (apart from a small amount which is used in making the materials employed in the construction of nuclear power stations) and because the rate at which it uses up resources of radioactive material is small compared to the total resource available. It is only efficiently generated in large units, so is suitable for supplying power to national grids or to large urban connurbations, but not for small, more localized supplies. The cost of nuclear energy compared with energy from fossil fuel sources (Fig. 11.13) is often a subject of debate; exactly where it falls in relation to the others depends on the return expected on the up-front capital cost, which represents a large element of the total. A further advantage of nuclear energy installations is that the technology is known; they can be built now and therefore contribute to the reduction of carbon dioxide emissions in the short term.

The continued importance of nuclear energy is recognized in the WEC energy scenarios, which all assume growth in this energy source next century. How much growth will be realized will depend to a large degree on how well the nuclear industry is able to satisfy the general public of the safety of its operations; in particular that the risk of accidents from new installations is negligible, that nuclear waste can be safely disposed of, that the distribution of dangerous nuclear material can be

Agenda 21 and energy

Agenda 21 is a document of some 400 or 500 pages accepted by the participating nations at the United Nations Conference on Environment and Development at Rio de Janeiro in June 1992. It covers a very wide range of topics concerned with the environment and development. The following paragraph 9.12 is contained in the chapter dealing with protection of the atmosphere:

'Governments ... with the cooperation of the relevant United Nations bodies and, as appropriate, intergovernmental and non-governmental organizations and the

private sector, should: (a) cooperate in identifying and developing economically viable, and environmentally sound energy sources to promote the availability of increased energy supplies to support sustainable development efforts, in particular in developing countries; ... (d) promote the research, development, transfer and use of technologies and practices for environmentally sound energy systems, including new and renewable energy systems, with particular attention to developing countries...'

---

effectively controlled and that it can be prevented from getting into the wrong hands.

A further potential nuclear energy source depends on fusion rather than fission. When, at extremely high temperatures, the nuclei of hydrogen (or one of its isotopes) are fused to form helium, a large amount of energy is released. This is the energy source which powers the sun; if a suitable method for harnessing and containing the process on the Earth could be devised, virtually limitless supplies of energy could be provided. A great deal of research is being put into fusion technology, but the economic generation of power by this means still appears to be many decades away.

Summary

This chapter has outlined the ways in which energy for human life and industry is currently provided. Growth in conventional energy sources at the rate required to meet future energy needs will generate much increased emissions of greenhouse gases. The WEC have

taken a lead by proposing a future energy scenario (scenario C) driven strongly by environmental considerations which would lead to the stabilization of carbon dioxide concentration in the atmosphere by the end of next century.

The WEC admits that the achievement of such a scenario (or a similar one which would lead to stabilization of carbon dioxide concentrations) will not be easy and will require both clear policies on the part of governments and cooperation from individuals. What is required now to begin to follow that pathway, however, is neither particularly difficult or costly. It just requires clear resolve. Four areas of action are important.

■ Many studies have shown that in most developed countries energy savings of 20 or 30 per cent can be achieved at no net cost or even at some overall saving. But industry and individuals will require not just encouragement, but modest incentives if the savings are to be realized.

■ Renewable energy sources (especially 'modern' biomass, wind and solar energy), which can begin to replace energy from fossil fuels, must be developed. For this to be done on an adequate scale, economic incentives will need to be applied. One means of providing an incentive could be by adding a component to the cost of fossil fuels which recognizes the environmental cost associated with their use

■ Arrangements are needed to ensure that technology is available for developing countries to develop their energy plans with high efficiency and to deploy renewable energy sources (for instance local solar energy generators) as widely as possible.

■ With world investment in the energy industry running at around one million million US dollars per year, there is a great responsibility on both governments and industry to ensure that these investments take long-term environmental requirements fully into account.

At the United Nations Conference on Environment and Development at Rio de Janeiro in June 1992, the countries of the world committed themselves in Agenda 21 (see box) to the action necessary to address the problems of energy and the environment. But the WEC are not optimistic that the necessary commitment exists to meet what is required, for instance to deliver their scenario C. They point out that 'the real challenge is to communicate the reality that the switch to alternative forms of supply will take many decades, and thus the realization of the need, and commencement of the appropriate action, must be *now*' (their italics)[50].

FOOTNOTES

1 Adapted from G.R. Davis, 'Energy for Planet Earth', *Sci. Amer.*, **263**, September 1990, pp 21–27.

2 1 toe = 11.7 MWh; 1 toe per day = 487kW.

3 *Energy for tomorrow's world—the realities, the real options and the agenda for achievement*, WEC Commission Report, World Energy Council 1993, p 41.

4 World Energy Council 1993, *op. cit.*

5 World Energy Council 1993, *op. cit.*

6 World Energy Council 1993 *op. cit.*

7 From World Energy Council 1993, *op. cit.*, p 27.

8 From World Energy Council 1993, *op. cit.*, p 304.

9 From World Energy Council 1993, *op. cit.*, p 28.

10 From World Energy Council 1993, *op. cit.*, p 29.

11 From UK government estimates 1993.

12 From World Energy Council 1993, *op. cit.*, p 122.

13 From World Energy Council 1993, *op. cit.*, p 113.

14 From National Academy of Sciences, *Policy Implications of Greenhouse Warming,* National Academy Press, Washington DC, 1992, Chapter 21.

15 From *Policy Implications of Greenhouse Warming,* pp 211, 212.

16 World Energy Council 1993, *op. cit.*, p 53 (source: Society of Motor Manufacturers and Traders UK).

17 World Energy Council 1993, *op. cit.*, p 53.

18 Example quoted in *Energy, environment and profits,* published by the Energy Efficiency Office of the Department of the Environment, UK 1993.

19 See for instance *Policy Implications of Greenhouse Warming,* Chapter 22, and World Energy Council 1993 *op. cit.*, Chapter 4.

20 World Energy Council 1993, *op. cit.*, p 94.

21 'Air Pollution Report No 66', *The ENDS Report*, **No 219**, April 1993, Environmental Data Services Ltd., London, pp 17–19.

22 J. Twidell and T. Weir, *Renewable Energy Resources*, E. & F.N. Spon, 1986, p 291.

23 'Large' hydro applies to schemes greater than 10 megawatts in capacity, 'small' hydro to schemes smaller than 10 megawatts.

24 From *Report of the Renewable Energy Advisory Group, Energy Paper Number 60*, UK Department of Trade and Industry, November 1992.

25 Two sources of comprehensive information about renewable energy are *Renewable Energy*, eds T.B. Johansson *et al.*, Island Press, Washington DC, 1993, and *World Energy Council Report:Renewable Energy Resources*, World Energy Council, London, 1993.

26 J.R. Moreira and A.D. Poole, 'Hydropower and its constraints' in Johansson *et al.* (eds) 1993 *op. cit.*, chapter 2, pp 73–119.

27 Further details in *World Energy Council Report*, 1993, *op. cit.*, section 7.

28 J.R. Moreira and A.D. Poole, *op. cit.*

29 From *Report of the Renewable Energy Advisory Group*, 1992, *loc. cit.*

30 See E. Mills, D. Wilson & T. Johansson in *Climate Change: Science, Impacts and Policy, Proceedings of the Second World Climate Conference*, eds J. Jager and H.L. Ferguson, CUP, 1991, pp 311–28.

31 From J. Twidell and A. Weir, *op. cit.*

32 See Royal Commission on Environmental Pollution, 17th Report, *Incineration of Waste*, HMSO, London, 1993, pp 43–47.

33 D.O. Hall *et al.*, 'Biomass for energy: supply prospects', in T.B. Johansson *et al.* 1993, *op. cit.*, pp 593–651.

34 From *Report of the Renewable Energy Advisory Group*, 1992, *loc. cit.*, pA29.

35 From *Report of the Renewable Energy Advisory Group*, 1992, *loc. cit.*

36 J. Twidell and A. Weir, *op. cit.*, p 252.

37 Adapted from J. Twidell and A.Weir, *op. cit.*, p 100.

38 J. Twidell and C. Johnstone, 'Glasgow gains from Strathclyde's solar residences', *Sun at work in Europe*, **Vol 7 No 4**, December 1992, pp 15–17.

39 N. Williams, K. Jacobson, H. Burris, 'Sunshine for light in the night', *Nature*, **362**, 1993, pp 691–92.

40 D. Lovejoy, 'Photovoltaics for rural electrification' in *Renewable Energy: Technology and the Environment*, ed. A.A.M. Sayigh, **Volume 1**, Pergamon Press, 1992, pp 124–32.

41 This figure assumes that about one third of total solar energy in 2020 will be PV electricity.

42 H. Kelly, 'Introduction to photovoltaic technology', in T.B.Johansson *et al.*, 1993 *op. cit.*, pp 297–336.

43 D.E. Carlson and S. Wagner, 'Amorphous silicon photovoltaic systems' in T.B. Johansson *et al.*, 1993 *op. cit.*, pp 403–436.

44 K. Zweibel and A.M. Barnett, 'Polycrystalline thin-film photovoltaics' in T.B. Johansson *et al.*, 1993 *op. cit.*, pp 437–82

45 World Energy Council Report 1993, *loc. cit.*, pp 2–48.

46 Adapted from National Academy of Sciences 1992, *op. cit.*, p 355.

47 T.B. Johansson *et al.*, 1993 *op. cit.*, p38.

48 From J.M. Ogden and J. Nitsch, 'Solar Hydrogen' in T.B. Johansson *et al.* (eds) 1993 *op. cit.*, pp 925–1009.

49 For more details of this technology see Ogden and Nitsch, *loc. cit.*

50 World Energy Council 1993, *op. cit.*, p 88.

12 *The Global Village*

*T*he preceding chapters have considered the various strands of the global warming story and the action that should be taken. In this last chapter I want first to present some of the challenges of global warming, especially those which arise because of its global nature. I then want to put global warming in the context of other major global problems faced by humankind.

The challenges of global warming

We have noted in the course of our discussion that global warming is not the only environmental problem. For instance, coastal regions are liable to subsidence for other reasons; water supplies in many places are already being depleted faster than they are being replenished and agricultural land is being lost through soil erosion. Many other reasons, locally and regionally, could be listed for the occurrence of environmental degradation. However, the importance of global warming is not diminished by the existence of these other environmental problems; in fact their existence will generally exacerbate its effects—and it is often necessary to tackle all related environmental problems together.

Local degradation of the environment is generally the result of particular action in the locality. To give an example, subsidence occurs because of the over-extraction of groundwater. In these cases the community where the malpractice is occurring suffers the damage which it causes and the principle that polluters should pay the cost of their pollution is relatively easy to apply.

That the 'polluter should pay' when the pollution is global rather than local is one of the Principles (Principle 16) enshrined in the Rio Declaration of June 1992. Appropriate mechanisms are needed to apply this principle on a global scale. The particular characteristic of global warming, compared with most environmental problems, is that it is *global*. However, though everybody contributes to it, to a greater or lesser extent, its adverse impact will not fall uniformly. Many, especially in the developing world, will experience significant damage; others will in fact gain from it. But the appreciation that an individual burning fossil fuels anywhere in the world has impact globally demands that a global attitude must be presented to the problem. It means also that global mechanisms must be generated for dealing with the problem.

There is already some experience in tackling an environmental problem of global scale: the depletion of stratospheric ozone because of the injection by humans of chlorofluorocarbons (CFCs) into the atmosphere possesses similar global characteristics to the global warming problem. An effective mechanism for tackling and solving the problem of ozone depletion has been established through the Montreal Protocol. All nations contributing to the problem have agreed to phase out their emissions of harmful substances. The richer nations involved have also agreed to provide finance and technology transfer to assist developing countries to comply. A way forward for addressing global environmental problems has therefore been charted.

Moving in that direction in the case of global warming will not be easy because the problem is so much larger and because it strikes so much nearer to the core of human resources and activities—such as energy and transport—upon which our quality of life depends. However, there are particular responsibilities and challenges for different communities of expertise which generally transcend national boundaries.

■ For the world's scientists the brief is clear: to provide better information especially about the expected climate change on the regional and local level, always keeping an appropriate emphasis on the uncertainties of prediction. Not only politicians and policy makers but also ordinary people need the information provided in the clearest possible form, in all countries and at all levels of society.

■ In the world of politics, it is well over a decade since Sir Crispin Tickell drew attention to the need for international action addressing

climate change[1]. Since then, a great deal of progress has been made with the signing in Rio in 1992 of the Climate Convention and with the setting up of the Sustainable Development Commission in the United Nations. The challenges presented to the politicians and decision makers by the Convention are first, to achieve the right balance of development against environmental concern, that is to achieve *sustainable* development, and secondly, to find the resolve to turn the many fine words of the Convention into genuine action.

■ As I have described the likely impacts of global warming and the ways in which it can be alleviated, the role of technology is clear. Technical challenges need to be picked up by the world's industry. Too often environmental concerns and environmental regulation are seen by industry as a threat, when in fact, they are an opportunity. Because of increasing public awareness of the environment and of the need for its preservation, the industries which are likely to grow and flourish next century are those which have taken environmental considerations firmly on board.

■ The responsibilities of industry (often working together with governments) must also be seen in the world context. The need to transfer appropriate technology to developing countries, especially in the energy sector, is particularly important. This has been specifically recognized in the Climate Convention which in Article 4, paragraph 5 states: 'The developed country Parties ... shall take all practical steps to promote, facilitate and finance, as appropriate, the transfer of, or access to, environmentally sound technologies and know-how to other Parties, particularly developing country Parties, to enable them to implement the provisions of the Convention.'

■ There are also new challenges for economists: for instance, the challenges of establishing environmental costs and of finding ways to represent the value of 'natural' capital, especially when it is of a global kind, mentioned in Chapter 9. There is the further problem of dealing fairly with all countries. No country wants to be put at a disadvantage economically because it has taken its responsibilities with respect to global warming more seriously than others. As economic instruments (for instance, taxes, subsidies, grants or other measures) are devised to provide the incentives for appropriate action regarding global warming by governments or by individuals, these must be seen to be both fair and effective for all nations. Economists working with politicians and decision makers need to find imaginative solutions which recognize not just environmental concerns but political realities.

Finally, it is important to recognize that the problem is not only global but long-term—the time-scales of climate change and of the influence of any action on programmes such as those of forestry, energy generation or transport are of the order of decades. The programme of action must therefore be seen as an evolving one, based on the continuing scientific, technical and economic assessments. Because it is evolving it is likely to require continual adjustment in the light, for instance, of new scientific knowledge or technical developments. But the long-term view must still be maintained.

Not the only global problem

Global warming is not the only global problem. There are other issues of a global scale and we need to see global warming in their context. Four problems of particular importance impact on the global warming issue.

Poverty and population growth

Prince Charles, in addressing the World Commission on Environment and Development on 22 April 1992, spoke as follows[2]:

'I do not want to add to the controversy over cause and effect with respect to the Third World's problems. Suffice it to say that I don't, in all logic, see how any society can improve its lot when population growth regularly exceeds economic growth. The factors which will reduce population growth are, by now, easily identified: a standard of health care that makes family planning viable, increased female literacy, reduced infant mortality and access to clean water. Achieving them, of course, is more difficult—but perhaps two simple truths need to be writ large over the portals of every international gathering about the environment: we will not slow the birth rate until we address poverty. And we will not protect the environment until we address the issue of poverty and population growth in the same breath.'

The first is population growth. When I was born there were about 2,000 million people in the world. By the end of the century there will be close to 6,000 million. During the lifetime of my grandchildren it is likely to rise to at least 10,000 million. Most of the growth will be in developing countries; by 2020 they will contain nearly 85 per cent of the world's people. These new people will all make demands for food, energy and work to generate the means of livelihood—all with associated implications for global warming.

The second issue is that of poverty and the increasing disparity in wealth between the developed and the developing world. The gap between the rich nations and the poor nations is becoming wider. Prince Charles[3] has drawn attention to the strong links which exist between population growth, poverty and environmental degradation (see box).

The third global issue is that of the consumption of resources, which in many cases is contributing to the problem of global warming. Many of the resources now being used cannot be replaced, yet we are using them at an unsustainable rate. In other words, because of the rate at which we are depleting them, we are seriously affecting their use even at a modest level by future generations. Sustainable development is all about the move to a situation in which this is no longer the case.

The fourth issue is that of global security. Our traditional understanding of security is based on the concept of the sovereign state with secure borders against the outside world. But communications, industry and commerce increasingly ignore state borders, and problems like that of global warming and the other global issues we have mentioned transcend national boundaries. Security therefore also needs to take on more of a global dimension.

The impacts of climate change may well pose a threat to security. One of the most recent wars has been fought over oil. It has been suggested

Al Gore's five strategic goals

Al Gore, the Vice-President of the United States, has proposed a plan for saving the world's environment[5]. He has called it 'A Global Marshall Plan' paralleled after the Marshall Plan through which the United States assisted western Europe to recover and rebuild after the Second World War. Resources for the plan would need to come from the world's major wealthy countries. He has proposed five strategic goals for the plan:

■ The stabilization of world population,

■ The rapid creation and development of environmentally appropriate technologies,

■ A comprehensive and ubiquitous change in the economic 'rules of the road' by which we measure the impact of our decisions on the environment.

■ The negotiation and approval of a new generation of international agreements; agreements which must be especially sensitive to the vast differences of capability and need between developed and developing nations.

■ The establishment of a cooperative plan for educating the world's citizens about our global environment.

that wars of the future could be fought over water[4]. The threat of conflict must be greater if nations lose scarce water supplies or the means of livelihood as a result of climate change. A dangerous level of tension could easily arise, with large numbers of environmental refugees. As Admiral Sir Julian Oswald has recently pointed out[6], a broader strategy regarding security needs to be developed which considers *inter alia* environmental threats as a possible source of conflict. In addressing the appropriate action to combat such threats, it may be better overall and more cost-effective in security terms to allocate resources to the removal or the alleviation of the environmental threat rather than to military or other measures to deal head-on with the security problem itself.

The goal of environmental stewardship

In the western world there are many material goals: economic growth, social welfare, better transport, more leisure and so on. But for our fulfilment as human beings we desperately need not just material challenges, but challenges of a moral or spiritual kind. There are strong connections, which I drew out in Chapter 8, between our basic attitudes, including religious belief, and environmental concern. I drew a picture of humans as stewards or gardeners of the Earth. Many people in the world are already deeply involved in a host of ways in matters of environmental concern. Such concern could, however, with benefit to us all, be elevated to a higher public and political level. Al Gore, the Vice-President of the United States, has suggested (see box) that we should embrace the preservation of the Earth as our new organizing principle. The United Nations, so far as it is able, has laid out a course of action. An appropriate challenge for everybody, from individuals, communities, industries and governments through to multinationals, especially for those in the relatively affluent Western world, is to take on board thoroughly this urgent task of the environmental stewardship of our Earth.

What the individual can do

I have spelled out the responsibilities of scientists, economists, politicians and industry. There are important contributions also to be made by ordinary individuals to help to mitigate the problem of global warming. Some of these are to:

■ ensure maximum energy efficiency in the home—through good insulation (see box in Chapter 11) against cold in winter and heat in summer, and through the use of appliances with high energy efficiency (see box in Chapter 11);

■ ensure maximum energy saving, for instance by making sure that rooms are not overheated and that light is not wasted;

■ drive a fuel-efficient car and choose means of transport which tend to minimize overall energy use;

■ check, when buying wood products, that they originate from a renewable source;

■ through the democratic process, encourage local and national governments to deliver policies which properly take the environment into account.

FOOTNOTES

1 Crispin Tickell, *Climatic Change and World Affairs*, Harvard University Press, second edition, 1986.

2 HRH the Prince of Wales in the First Brundtland Speech, 22 April 1992, published in *Threats without enemies*, ed. G.Prins, Earthscan, 1993, pp 3–14.

3 HRH the Prince of Wales, *loc. cit.*, 1993.

4 The United Nations Secretary-General, Boutros Boutros-Ghali, has said that 'the next war in the Middle East will be fought over water, not politics'.

5 Expounded in the last chapter of Al Gore, *Earth in the balance*, Houghton Miflin Company, 1992.

6 Admiral Sir Julian Oswald, 'Defence and Environmental Security' in G. Prins (ed.) *op. cit.*, 1993, pp 113–34.

GLOSSARY

Agenda 21 A document accepted by the participating nations at *UNCED* on a wide range of environmental and development issues

Albedo The fraction of light reflected by a surface, often expressed as a percentage. Snow-covered surfaces have a high albedo level; vegetation-covered surfaces a low albedo, because of the light absorbed for *photosynthesis*

Atmosphere The envelope of gases surrounding the Earth or other planets

Atmospheric pressure The pressure of atmospheric gases on the surface of the planet. High atmospheric pressure generally leads to stable weather conditions, whereas low atmospheric pressure leads to storms such as *cyclones*

Atom The smallest unit of an *element* that can take part in a chemical reaction. Composed of a nucleus which contains *protons* and *neutrons* and is surrounded by *electrons*

Atomic mass The sum of the numbers of *protons* and *neutrons* in the nucleus of an *atom*

Biodiversity A measure of the number of different biological species found in a particular area

Biological pump The process whereby carbon dioxide in the *atmosphere* is dissolved in sea water where it is used for *photosynthesis* by *phytoplankton* which are eaten by *zooplankton*. The remains of these microscopic organisms sink to the ocean bed, thus removing the carbon from the *carbon cycle* for hundreds, thousands or millions of years

Biomass The total weight of living material in a given area

Biome A distinctive ecological system, characterized primarily by the nature of its vegetation

Biosphere The region on land, in the oceans and in the *atmosphere* inhabited by living organisms

Business-as-usual The scenario for future world patterns of energy consumption and *greenhouse gas* emissions which assumes that there will be no major changes in attitudes and priorities

C3, C4 plants Groups of plants which take up *carbon dioxide* in different ways in *photosynthesis* and are hence affected to a different extent by increased *atmospheric* carbon dioxide. Wheat, rice and soya bean are C3 plants; maize, sugarcane and millet are C4 plants

Carbon cycle The exchange of carbon between the *atmosphere*, the land and the seas

Carbon dioxide One of the major *greenhouse gases*. Human-generated carbon dioxide is caused mainly by the burning of *fossil fuels* and *deforestation*

Carbon dioxide fertilization effect The process whereby plants grow more rapidly under an *atmosphere* of increased *carbon dioxide* concentration. It affects *C3 plants* more than *C4 plants*

CEE Communité Economique Europaen (European Economic Community)

Celsius Temperature scale, sometimes known as the Centigrade scale. Its fixed points are the freezing point of water (0°C) and the boiling point of water (100°C)

CFCs Chlorofluorocarbons; synthetic compounds used extensively for refrigeration and aerosol sprays until it was realized that they destroy ozone (they are also very powerful *greenhouse gases*) and have a very long lifetime once in the *atmosphere*. The Montreal Protocol agreement of 1987 will result in the scaling down of CFC production and use in industrialized countries

Chaos A mathematical theory describing systems which are very sensitive to the way they are originally set up; small discrepancies in the initial conditions will lead to completely different outcomes when the system has been in operation for a while. For example, the motion of a pendulum when its point of suspension undergoes forced oscillation will form a particular pattern as it swings. Started from a slightly different position, it can form a completely different pattern, which could not have been predicted by studying the first one. The weather is a partly chaotic system, which means that even with perfectly accurate forecasting techniques, there will always be a limit to the length of time ahead that a useful forecast can be made

CIS Commonwealth of Independent States (former USSR)

Climate The average weather in a particular region

Climate sensitivity The global average temperature rise under doubled *carbon dioxide* concentration in the *atmosphere*

Compound A substance formed from two or more *elements* chemically combined in fixed proportions

Condensation The process of changing state from gas to liquid

Convection The transfer of heat within a fluid generated by a temperature difference

Coppicing Cropping of wood by judicious pruning so that the trees are not cut down entirely and can regrow

Tropical cyclone A storm or wind system rotating around a central area of low *atmospheric pressure*. They can be of great strength and are also called hurricanes and typhoons. Tornadoes are much smaller storms of similar violence

Daisyworld A model of biological *feedback* mechanisms developed by James Lovelock (see also *Gaia hypothesis*)

DC Developing country—also Third World country

Deforestation Cutting down forests; one of the causes of the enhanced *greenhouse effect*, not only when the wood is burned or decomposes, releasing *carbon dioxide*, but also because the trees previously took carbon dioxide from the *atmosphere* in the process of *photosynthesis*

Deuterium Heavy *isotope* of hydrogen

Drylands Areas of the world where precipitation is low and where rainfall consists of small, erratic, short, high-intensity storms

Ecosystem A distinct system of interdependent plants and animals, together with their physical environment

El Nino A pattern of ocean surface temperature in the Pacific off the coast of South America, which has a large influence on world *climate*

Electron Negatively charged component of the *atom*

Element Any substance that cannot be separated by chemical means into two or more simpler substances

Environmental refugees People forced to leave their homes because of environmental factors such as drought, floods, sea-level rise

Evaporation The process of changing state from liquid to gas

FAO The United Nations Food and Agriculture Organization

Feedbacks Factors which tend to increase the rate of a process (positive feedbacks) or decrease it (negative feedbacks), and are themselves affected in such a way as to continue the feedback process. One example of a positive feedback is snow falling on the Earth's surface, which gives a high *albedo* level. The high level of reflected rather than absorbed *solar radiation* will make the Earth's surface colder than it would otherwise have been. This will encourage more snow to fall, and so the process continues

Fossil fuels Fuels such as coal, oil and gas made by decomposition of ancient animal and plant remains which give off *carbon dioxide* when burned

Gaia hypothesis The idea, developed by James Lovelock, that the *biosphere* is an entity capable of keeping the planet healthy by controlling the physical and chemical environment

Geoengineering The artificial modification of the *atmosphere* to counteract *global warming*

Geothermal energy Energy obtained by the transfer of heat to the surface of the Earth from its depths

Global warming The idea that increased *greenhouse gases* cause the Earth's temperature to rise globally (see *greenhouse effect*)

Green Revolution Development of new strains of many crops in the 1960s which increased food production dramatically

Greenhouse effect The cause of *global warming*. Incoming *solar radiation* is transmitted by the *atmosphere* to the Earth's surface, which it warms. The energy is retransmitted as *thermal radiation*, but some of it is absorbed by *molecules* of greenhouse gases instead of being retransmitted out to space, thus warming the atmosphere. The name comes from the ability of greenhouse glass to transmit incoming solar radiation but retain some of the outgoing thermal radiation to warm the interior of the greenhouse. The 'natural' greenhouse effect is due to the greenhouse gases present for natural reasons, and is also

observed for the neighbouring planets in the solar system. The 'enhanced' greenhouse effect is the added effect caused by the *greenhouse gases* present in the atmosphere due to human activities, such as the burning of *fossil fuels* and *deforestation*

Greenhouse gases *Molecules* in the Earth's *atmosphere* such as *carbon dioxide* (CO_2), methane (CH_4) and *CFCs* which warm the atmosphere because they absorb some of the *thermal radiation* emitted from the earth's surface (see the *greenhouse effect*)

Greenhouse gas emissions The release of *greenhouse gases* into the *atmosphere*, causing *global warming*

GtC Gigatonnes of carbon (C) (1 gigatonne = 10^9 tonnes). 1 GtC = 3.7 Gt *carbon dioxide*

GWP Global warming potential: the ratio of the enhanced *greenhouse effect* of any gas compared with that of *carbon dioxide*

Heat capacity The amount of heat input required to change the temperature of a substance by 1°C. Water has a high heat capacity so it takes a large amount of heat input to give it a small rise in temperature

Hydrological (water) cycle The exchange of water between the *atmosphere*, the land and the seas

Hydro-power The use of water-power to generate electricity

IPCC Intergovernmental Panel on Climate Change—the world scientific body assessing *global warming*

Isotopes Different forms of an *element* with different *atomic masses*; an element is defined by the number of *protons* its nucleus contains, but the number of *neutrons* may vary, giving different isotopes. For example, the nucleus of a carbon atom contains six protons. The most common isotope of carbon is ^{12}C, with six neutrons making up an atomic mass of 12. One of the other isotopes is ^{14}C, with eight neutrons, giving an atomic mass of 14. Carbon-containing *compounds* such as *carbon dioxide* will contain a mixture of ^{12}C and ^{14}C isotopes. See also *deuterium, tritium*

Latent heat The heat absorbed when a substance changes from liquid to gas (*evaporation*), for example when water evaporates from the sea surface using the sun's energy. It is given out when a substance changes from gas to liquid (*condensation*), for example when clouds are formed in the *atmosphere*

Milankovitch forcing The imposition of regularity on *climate* change triggered by regular changes in

distribution of *solar radiation* (see *Milankovitch theory*)

Milankovitch theory The idea that major ice ages of the past may be linked with regular variations in the Earth's orbit around the sun, leading to varying distribution of incoming *solar radiation*

MINK Region of the United States comprising the states of Missouri, Iowa, Nebraska and Kansas, used for a detailed *climate* study by the US Department of Energy

Molecule Two or more *atoms* of one or more *elements* chemically combined in fixed proportions. For example, atoms of the elements carbon (C) and oxygen (O) are chemically bonded in the proportion one to two to make molecules of the *compound carbon dioxide* (CO_2). Molecules can also be formed of a single element, for example ozone (O_3)

Monsoon Particular seasonal weather patterns in sub-tropical regions which are connected with periods of heavy rainfall

Neutron A component of most atomic nuclei without electric charge, of approximately the same mass as the *proton*

OECD Organization for Economic Cooperation and Development

Ozone hole A region of the *atmosphere* over Antarctica where, during spring in the southern hemisphere, about half the atmospheric ozone disappears

Paleoclimatology The reconstruction of ancient *climates* by such means as ice-core measurements. These use the ratios of different *isotopes* of oxygen in different samples taken from a deep ice 'core' to determine the temperature in the *atmosphere* when the sample *condensed* as snow in the clouds. The deeper the origin of the sample, the longer ago the snow became ice (compressed under the weight of more snowfall)

Passive solar design The design of buildings to maximize use of *solar radiation*. A wall designed as a passive solar energy collector is called a solar wall

Photosynthesis The series of chemical reactions by which plants take in the sun's energy, *carbon dioxide* and water vapour to form materials for growth, and give out oxygen. Anaerobic photosynthesis takes place in the absence of oxygen

Phytoplankton Minute forms of plant life in the oceans

ppbv parts per thousand million by volume (measurement of concentration)

ppmv parts per million by volume (measurement of concentration)

Precautionary Principle The principle of prevention being better than cure, applied to potential environmental degradation

Primary energy Energy sources, such as *fossil fuels*, nuclear or wind power, which are not used directly for energy but transformed into light, useful heat, motor power and so on. For example, a coal-fired power station which generates electricity uses coal as its primary energy

Proton A positively charged component of the atomic nucleus

PV Photovoltaic: a solar cell often made of silicon which converts *solar radiation* into electricity

Radiation budget The breakdown of the radiation which enters and leaves the Earth's *atmosphere*. The quantity of *solar radiation* entering the atmosphere from space should be balanced by the *thermal radiation* leaving the Earth's surface and the atmosphere

Radiative forcing (a) The overall effect, due to any specified gas, on the amount of *thermal radiation* from the Earth leaving the top of the *atmosphere* (b) Cloud radiative forcing is the overall effect of a cloud on the thermal radiation leaving the top of the atmosphere

Renewable energy Energy sources which are not depleted by use, for example *hydro-power*, *PV* solar cells, wind power and *coppicing*

Respiration The series of chemical reactions by which plants and animals break down stored foods with the use of oxygen to give energy, *carbon dioxide* and water vapour

Sequestration Removal and storage, for example, *carbon dioxide* taken from the *atmosphere* into plants via *photosynthesis*

Solar radiation Energy from the sun

Sonde A device sent up to obtain information such as temperature and *atmospheric pressure* about *atmospheric* conditions, for example radio or balloon sondes

Stewardship The attitude that human beings should see the Earth as a garden to be cultivated rather than a treasury to be raided. (See also *sustainable development*)

Sustainable development Development which meets the needs of the present without compromising the ability of future generations to meet their own needs

Thermal radiation Radiation emitted by all bodies, in amounts depending on their temperature. Hot bodies emit more radiation than cold ones

Transpiration The transfer of water from plants to the *atmosphere*

Tritium Radioactive *isotope* of hydrogen, used to trace the spread of radioactivity in the ocean after atomic bomb tests, and hence to map ocean currents

Watt Unit of power

UNCED United Nations Conference on Environment and Development, held at Rio de Janeiro in June 1992, after which the United Nations Framework Convention on Climate Change was signed by 160 participating countries

UNEP United Nations Environmental Programme—one of the bodies which set up the *IPCC*

UV Ultra-violet radiation

WEC World Energy Council—an international body with a broad membership of both energy users and the energy industry

Wind farm Grouping of wind turbines for generating electric power

WMO World Meteorological Organization—one of the bodies which set up the *IPCC*

Younger Dryas event Cold climatic event which occurred for a period of about 1,500 years, interrupting the warming of the Earth after the last ice age (so called because it was marked by the spread of an Arctic flower, the 'Dryas octopetala'). It was discovered by a study of *paleoclimatic* data

Zooplankton Minute forms of animal life in the ocean

Chemical symbols

CFCs	chlorofluorocarbons
CH_4	methane
CO	carbon monoxide
CO_2	carbon dioxide
H_2	molecular hydrogen
HCFCs	hydrochlorofluorocarbons
H_2O	water
N_2	molecular nitrogen
N_2O	nitrous oxide
NO	nitric oxide
NO_2	nitrogen dioxide
O_2	molecular oxygen
O_3	ozone
OH	hydroxyl radical
SO_2	sulphur dioxide

Unit abbreviations

Quantity	Prefix	Symbol
10^{12}	tera	T
10^9	giga	G
10^6	mega	M
10^3	kilo	k
10^2	hecto	h
10	deka	d
10^{-1}	deci	d
10^{-2}	centi	c
10^{-3}	milli	m
10^{-6}	micro	μ
10^{-9}	nano	n

INDEX

188